AR-15

RECOIL Magazine's Collection of Unique and Exceptional ARs

COMPILED BY
THE EDITORS OF
RECOIL MAGAZINE

Published by

RECOIL

Recoil Books, an imprint of Caribou Media Group, LLC
5600 W. Grande Market Drive, Suite 100
Appleton, WI 54913
recoilweb.com | www.gundigest.com

To order books or other products call 920.471.4522 ext.104
or visit us online at www.gundigeststore.com

ISBN-13: 978-1-951115-53-1

Edited and designed by Recoil Staff

Printed in the United States of America

10 9 8 7 6 5 4 3 2 1

TABLE OF CONTENTS

Here it is, everything you could possibly want to know about America's favorite rifle, hot off the presses and dripping with ballistic goodness. Over the past decade, RECOIL has compiled a vast compendium of AR-15 articles, and we've collected them here so that they're all at your fingertips, ranging from the mundane to the weird, the sublime to the ridiculous. We delve into the basic operating principles for the platform, how to address its shortfalls, (because like everything else created by the hand of man, it's far from perfect), and how to adapt it to various roles and ranges, from a CQB blaster to thousand-yard precision rig.

There's a metric ton of ideas for you to build your own perfect long gun, with information gleaned from first-hand use in the field — what works, what doesn't, where things can be improved, and what we'd look for in our personal guns. There are a few builds based on the carbines issued to our staffers, and some that came about from experience gained from carrying those carbines into conflict, and we've built more than a few just for grins and because no one told us we couldn't.

In addition to ARs we've had a personal hand in, there are plenty of examples from leading manufacturers who have invested untold sums of R&D dollars to iron out any wrinkles in performance, so that you can pick something off the shelf to scratch a particular itch. We've never shied away from voicing an opinion on the strengths or weaknesses of a particular product, so you'll get the unvarnished truth to help make any decisions regarding whether to whip out a credit card.

Pull up a comfortable chair, pour yourself a beverage and settle in for some in-depth reading about the rifle that more than any other defines this great nation.

Iain Harrison
EDITOR-IN-CHIEF

1 Basic

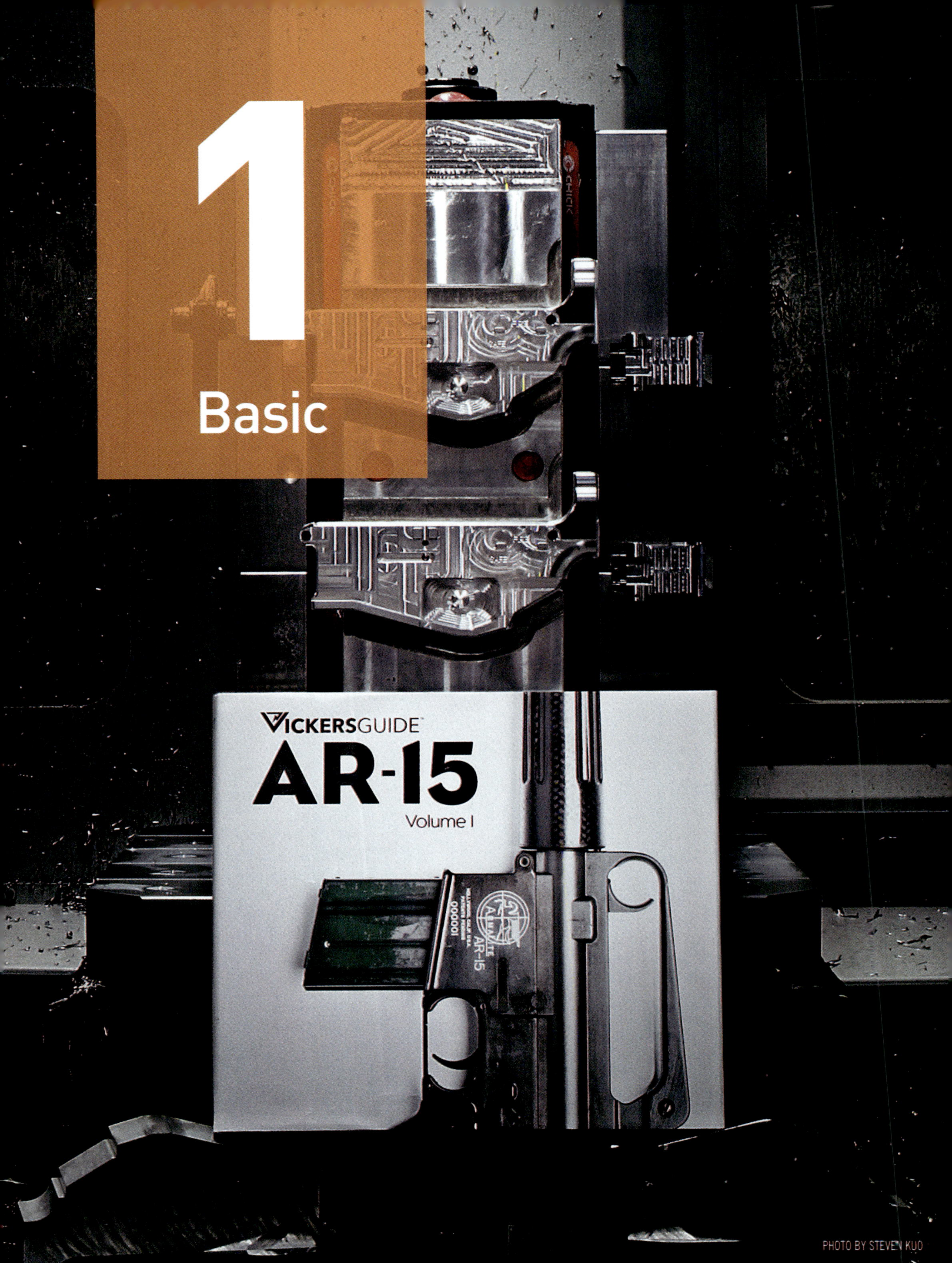

PHOTO BY STEVEN KUO

AMERICA'S RIFLE THE AR-15

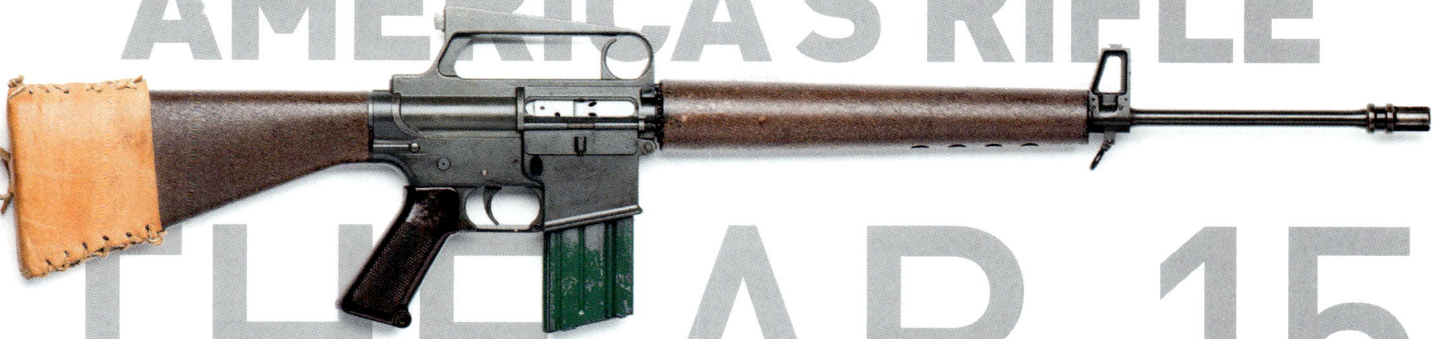

Story by Steven Kuo
with Commentary from Larry Vickers, C. Reed Knight Jr., and Dr. Earl Burress
Photos by James Rupley and Larry Vickers

The AR-15 has truly become "America's Rifle," modern-day muskets that are integral to the fabric of life for American gun owners. Whether wielded by warfighters and law enforcement, or in the hands of civilians for home defense, hunting, competition, or target shooting, ARs are literally everywhere. It's reported that one of every four rifles made in the U.S. is an AR. And it's a uniquely American story — with ingenuity and innovation, missteps, continuous improvements and refinements, and its eventual ascendance to firearms behemoth.

Today, one might think of Hollywood, California, with some amount of disdain when it comes to firearms. But back in the late 1950s, Armalite's small workshop in that very town gave birth to the most adaptable and longest-lasting standard U.S. service rifle of all time. The AR-15 was based on Eugene Stoner's 7.62mm AR-10 design, utilizing the latest in materials as well as manufacturing processes. Precise manufacturing tolerances combined with steel, aluminum, and synthetic materials to create a weapon system that has become a model of modularity, versatility, and effectiveness.

After a rough start in service during the Vietnam War, the U.S. military officially adopted the AR-15 platform as the M16A1 service rifle by the end of the decade. Chambered in 5.56x45mm, it's a direct impingement gas-operated rifle. Both of these facts have generated a great deal of debate and Internet butt-hurt over the years, but the truth is that the AR platform is extremely effective and has been refined to a mirror finish. There are many very lethal 5.56mm rounds on the market now, and the modularity of the platform allows you to easily switch to many other alternative calibers. The direct impingement operating system can deliver great accuracy and shootability, but many short- and long-stroke gas piston systems and conversions are also available. The AR platform features largely unparalleled ergonomics, with well-positioned controls, conveniently configured sighting systems, amazing aftermarket triggers, and the ability to accommodate different types and sizes of shooters. And there's an enormous ecosystem of products and services supporting the huge installed base of AR-platform rifles in the U.S. and the world. 'Merica, indeed.

Last year in Issue 22, we brought you a special look at some amazing 1911s courtesy of the *Vickers Guide: 1911*. If there's something even more closely associated with Larry Vickers than the 1911, it might just be the AR-15 platform. Vickers spent over two decades in special operations, including 15 years with Delta Force, deploying with different varieties of the AR platform. He's also renowned for his encyclopedic knowledge of firearms, gunsmithing, and firearms instruction.

Vickers Guide is back, with another beautiful coffee table book dedicated to the AR-15. With all the modern-day wonder guns, bristling with optics, accessories, and race parts, folks today might not be as familiar with early incarnations of the AR platform. So flip the page and enjoy some fascinating and historic rifles from the earliest days of the AR-15, with commentary from Vickers as well as industry experts C. Reed Knight Jr. of the Institute of Military Technology and Dr. Earl Burress of Tactical Applications Group. Please try not to drool on the magazine.

XAR1501

This is the first AR-15 ever made. We'll repeat that — this is the very first AR-15, ever. Patient zero. The granddaddy of them all. Built by Armalite before 1958,

Left: The first AR-15 ever made. Behold its glory. Note the charging handle on the top of the upper receiver and non-adjustable rear sight.

Right: One of the five original AR-10 prototypes. It was built in Armalite's workshop in Hollywood, California.

this is Experimental AR-15 Number 1, prior to Colt's involvement in the AR-15 program. On the orders of General Willard Wyman, Commanding General of the U.S. Continental Army Command, approximately 15 to 20 guns were built for initial testing.

Knight relates that the serial number, XAR1501, was derived from the following formula: eXperimentalAR15number01. Note that this particular rifle is missing the front sight and is equipped with a non-adjustable rear sight, as requested by General Wyman. The furniture was also painted with the same speckle paint used on the Armalite AR-10s.

If you look carefully at the left side of the rifle's magwell, you'll see something that's practically invisible to the naked eye in person. "AR-15 Exp. Prototype" is handwritten on the magwell. It all started here, folks.

"HOLLYWOOD" AR-10 SERIAL NUMBER 1003

Five original prototypes of the Armalite AR-10 were built, with serial numbers 1001 through 1005. The one shown here is the third, serial number 1003. The serial numbers for these initial guns were derived as follows: ar10number03. So number 1003 here is the third rifle built, not the 1,003rd.

Knight explains that they were all hand-built in Armalite's workshop in Hollywood, California, thus earning the moniker, "Hollywood AR-10s," and differentiating them from AR-10s later manufactured in other locations such as the Netherlands.

All five prototypes were inscribed with the large "ARMALITE" marking on the left side of the receiver. They were all used for demonstrations and originally had aluminum composite barrels. Serial number 1001 was sold to Sam Cummings at Interarms and remains with only minor modifications to its original configuration. Serial number 1002 was used in the Springfield Armory testing performed in 1957 and sports Mathewson's replacement barrel. Serial number 1003, shown here, was at Fairchild (Fairchild Engine and Airplane Corporation, the parent company of Armalite) and is virtually unaltered. Serial number 1004 was also tested at Springfield, fired extensively at the same time as number 1002, and later had the upper receiver replaced with a quick-change heavy barrel. 1005 was retrieved out of Europe, received additional modifications, and was the test rifle for the Dutch to build their original guns. These five guns were the very first batch of prototype hand-built guns with the unique-looking front sight tower.

COLT ACR

In the latter half of the 1980s, the U.S. Department of Defense funded the Advanced Combat Rifle program to develop a successor to the M16 platform. A primary objective of the program was to improve the hit probability of the average U.S. soldier employing the weapon, as compared to M16 variants in use at the time. Colt was one of six companies selected to participate, and the Colt ACR shown here (in its more rare carbine-length form) was their entry.

The Colt ACR experimented with a number of ergonomic changes to the M16 and is closely associated with the development of so-called "duplex" ammunition — a single cartridge containing two nested bullets, which would both be discharged down the barrel of the rifle with a single trigger pull and single recoil impulse, with the intent of increasing hit probability.

Vickers explains that the Colt ACR ultimately demonstrated that duplex ammunition fired from an M16 is nearly worthless due to unacceptable accuracy at reasonable combat distances. The effort to stabilize two relatively lightweight projectiles, each under 40 grains, out of a barrel designed to fire much heavier projectiles like the 62-grain M855 and 64-grain M856 tracer proved hopeless without a change of the barrel twist rate.

Additionally, the shotgun-style sighting plane on top of the handguards also wasn't sufficiently effective, with any such advantages as a rapid sighting system being completely outclassed by developments in red-dot optics to come later. As a point of comparison, Vickers himself used an Aimpoint 2000 optic on his Colt Model 723 during Operation Just Cause in December 1989 — the same time period that Colt and the U.S. Army were testing the ACR.

The program was eventually terminated in 1990, with none of the unique entrants progressing further. Vickers notes one particularly useful realization from the program — that an enhanced multi-position retractable buttstock would be advantageous. Indeed, the ACR-style buttstock lives on through a variety of manufacturers such as Lewis Machine & Tool, Knight's Armament Company, and B5. Alas, beyond that the Colt ACR has been relegated to history, with none of its other distinctive features making it past the developmental stage.

SON TAY COLT GAU-5A/A REPRODUCTION

The Son Tay raid, code-named Operation Ivory Coast, was a groundbreaking American special operations mission conducted to rescue U.S. prisoners of war at

A rare carbine-length Colt ACR, with enhanced buttstock and shotgun-style sighting plane. Pictured to the right is a cutaway of the unique but ill-fated duplex ammunition.

This is a reproduction of the Colt GAU-5A/A carbine used in Operation Ivory Coast, the Son Tay raid.

the Son Tay camp on the night of November 21, 1970.

Dr. Burress explains: 65 American prisoners were believed to be held at the Son Tay camp, located 21 miles from the North Vietnamese capital of Hanoi. During the operation, a team of 56 Army personnel were inserted by helicopter and assaulted the camp. In just 29 minutes, the assault team secured the objective, cleared the prison facility, determined that there were no American prisoners present, and completed an extraction of all U.S. personnel. Though they encountered heavy resistance, no Americans were killed, whereas the North Vietnamese forces suffered an estimated 40 killed-in-action. Although considered an intelligence failure since the American prisoners had already been moved out of the camp, the raid was an operational and tactical success. Moreover, it had a strategic impact on the Vietnam War, demonstrating America's will and capability to conduct a deep strike into North Vietnam.

During the operation, American forces primarily used two versions of the GAU-5 (the Air Force designation for the Colt XM177), many of which were equipped with a Normark Singlepoint Occluded Eye Gunsight (OEG) scope. It's believed that Air Force GAU-5s were selected in lieu of Army XM177s because the mission was rehearsed near Eglin Air Force Base, Florida; thus, it may have been easier to procure Air Force weapons through the local logistics system. Though today these weapons would be considered carbines, during this period the GAU-5 and XM177 family were classified as submachine guns. The 10-inch barreled GAU-5A carbine was assembled using partial fence lowers, and the 11.5-inch barreled GAU-5A/A was constructed using full fence lowers. The lower was fitted with a slick side upper devoid of a forward assist, a barrel with a 1 in 12-inch twist rate, an XM177 moderator, and an aluminum buttstock on a two-position buffer tube.

This Model 601 may be one of the prettiest AR-15s ever made.

During the mission rehearsal period between September and November 1970, the rifles were fitted with OEG red-dot optics to improve the nighttime capabilities of the ground assault force. This model Singlepoint scope (as pictured) presented the shooter with a closed black tube punctured by a 16 MOA pinhole, illuminated by a red fiber optic filament, which appeared to the shooter as a single red-dot. The weapon was aimed using the shooter's binocular vision to superimpose the red-dot, viewed by the shooter's dominant eye, over the target as viewed by the shooter's non-dominant eye. The Singlepoint was attached to the upper receiver's carrying handle using a mount that incorporated steel strap retention bands, similar to those used on the German WWII G43 ZF4 scope mount — reportedly reinforced with black electrical tape. The Son Tay Raid was the first documented example of a U.S. SOF team using a red-dot type optic in combat.

The reproduction of the Colt GAU-5A/A used in 1970 shown on the preceding page was constructed using an early 1970s upper, a barrel and lower from a mid-1990s Colt R6520, and a solid replica flash suppressor serving as a substitute for an XM177 moderator. A genuine moderator is classified as a sound suppressor/silencer by the BATFE, despite its minimal sound reduction. The replica flash suppressor is constructed to appear to be fitted with a grenade ring, designed to be used as a mount for the experimental XM148 40mm Grenade Launcher. The barrel length is a bit longer than 11.5 inches to achieve a 16.1-inch overall length with the moderator permanently attached, avoiding classification as a short-barreled rifle.

**COLT AR-15 (MODEL 601)
SERIAL NUMBER 000112**

This Model 601, serial number 000112, features absolutely beautiful, unpainted bakelite furniture. This is very unusual, and the fact that this rifle is the 14th Colt AR-15 Model 601 ever made just adds to the appeal. When it comes to aesthetics, Vickers considers this rifle to be at the top of the list for ARs.

The scope attached to this rifle is very rare in the U.S.; it's an adaptation of the same Delft scope used on the Dutch-built AR-10s. Different versions of this optic were also used on the unique Dutch variant of the FN FAL, which served as the standard service rifle in the Netherlands throughout most of the Cold War, before being replaced by Diemaco-made AR-style rifles. **R**

Presented in a collectible 13x11-inch coffee table format, *Vickers Guide: AR-15 (Volume 1)* features over 70 rifles, including some of the rarest and most historically unique rifles in the AR family. Options will include a Standard Edition ($95), a signed Signature Edition ($125), and a Limited Edition Premium Version ($250), limited to just 250 copies. ➔ **www.vickersguide.com**

EDITOR'S NOTE: This article was written and photographed prior to Tim Dillon being named CEO of Noveske Rifleworks in early summer. To learn what Tim has planned for the company and for the most up-to-date information, please go to RECOILweb.com.

THE NEXT GENERATION

An Insider's Look at Noveske Rifleworks, Its Gen III Carbines, and the Legacy of Its Founder

By Richard Kim
Photography by Stickman

When John Noveske died tragically in an auto accident in January 2013, the news was shocking. The firearms community mourned. It also had people wondering what would become of his eponymous company, known for pushing the limits of quality, high-end carbines. Well, big things have happened at Noveske Rifleworks in the past 18 months or so, and it's putting out more AR-15s than ever, while still meeting (or even exceeding) the high standards John set when he founded the business in 2001.

His widow Lorina Noveske retains ownership of the company. Lorina is also an artist (a painter), but now finds herself focusing her creative endeavors more often on layout, packaging, design and merchandising than on canvas. Meanwhile, longtime Noveske employee Sheri Johnson continues her position at the helm. While she was promoted to general manager, anyone who has spent time around her is aware that she also handles a lot of the company's day-to-day operations, including all the phone calls, email, and non-stop text messages that go along with it. Sheri was the go-to person before John's death and is even more so now.

The more we spoke with Sheri, the more apparent it became: she *really* knows AR-15s and the industry. While she looks like a SHOT Show "booth girl," she can talk weapons for hours — and likes doing it. If that weren't enough, Sheri is a shooter and attends multiple shooting courses a year, both rifle and pistol. She showed us the inner workings of their headquarters and gave us a firsthand look at how the company has not only maintained John's vision of building superior-quality firearms, but also expanded on his legacy.

THE THIRD GENERATION

The Noveske crew is proud of their Gen III lineup of weapons. One of the most common questions people ask is, "Why do a Gen III instead of sticking with Gen II?" Sheri had an immediate answer for that one, stating that the Gen 2 platform had a substantial rejection rate for quality-control issues. She elaborated, explaining that the Noveske "Chainsaw" lineup was very popular, but that having a large amount of blemished items was a failure for any business no matter how you looked at it.

By bringing upper and lower receiver manufacturing in house, they were now able to control all aspects including finishes. The finish on the upper and lower of the Gen II receivers had always had some variation, but new anodizing processes, as well as in-house machine work, allowed for a perfect match with the Gen III. For anyone who hasn't done anodizing work, the surface finish, length of time in the vat, and strength of chemicals involved are just a few of the things that affect the outcome.

Another complaint voiced on Internet forums was how tight the pivot and takedown pins fit in some Gen II lower and upper combinations. Retaining control of all CNC machine work allows for a pushpin that isn't too tight or loose, and more importantly, that will be consistent with combinations of other Gen III receivers.

All Gen III receivers and rails are also finished with Cerakote and will be available to dealers and distributors in black, flat dark earth (FDE), and olive drab (OD) green. Minimum orders may apply, so one distributor, namely RSR, has already locked in the FDE Noveske Infidel (which is the carbine pictured in these pages) as an exclusive. Standard Noveske Infidel carbines are still available in black as well as OD green through participating dealers.

Gen III weapons share another feature introduced by Sheri — a steel barrel nut with three different handguard options. After the smooth-sided NSR handguard, the traditional quad rail was brought in-house next along with some minor changes, including the barrel nut. Next up is the Hybrid rail, which also shares the NSR barrel nut and features a traditional 1913 rail on the top and bottom while providing slick KeyMods along the sides. (The Hybrid rail was not yet released during the writing of this article, but had a tentative street date of around the time that this issue hits newsstands.)

BUILDING PRECISION

Chris Quigley walked us through the factory to show us machining of the various components in the Noveske Gen III weapons. We started in the barrel shop, where heavy-duty shelves filled with blanks awaited us. John Noveske explained during an earlier RECOIL interview that he had his barrel blanks made in the Pac-Nor factory. Pac-Nor employees clock out on Pac-Nor time and clock in on Noveske time. Noveske uses its own special blend of steel, with tooling such as the button, reamers, etc., made specifically for them. While it would be easy enough to say Noveske makes the blanks, John was always pretty open about this. John was also very open about his fondness for Pac-Nor.

The blanks were trued up, both from the side facing the chamber and off the center to avoid any lopsided or crooked barrels. Profiling was next, followed by chambering — handled by a fellow named David. We asked how many barrels a reamer was good for. David smiled and answered that there was no solid number. Some would work for hundreds of barrels, while others only a small amount — or in a few cases, only one barrel. There are a lot of barrel companies that use CNC machines for chambering, with reamers that are thrown out when they no longer meet tolerance. Though that handles the science, David goes to show there is more to barrels than just the numbers. It is hard to be sure if that puts David into the category of old-school craftsman or modern-day wizard, but either way, there's a reason why Noveske is known for its impressive barrels.

Next up was installation of the barrel extension. The barrel extension is the first thing a bullet sees as it exits the magazine and what locks to the bolt, making the weapon safe to fire. A barrel extension that comes loose is unsafe to fire, and one that breaks while being fired may result in a trip to the hospital. Each barrel extension is checked as it's unpacked and typically deburred. The barrel extension is first locked into a hydraulic assembly, then torqued into place. To ensure it will not come free, Rocksett is also applied before installation.

Only after the barrel extension is installed can the gas port be drilled; a special jig is used to ensure everything is properly aligned. Different barrel lengths, calibers, and gas systems are optimized with different gas port sizes. Make it too small and the AR-15 will have a harder time cycling, though it will be a very soft-shooting weapon. Hog out that gas port and it will eat the weakest ammo, all the while slowly beating itself to death with increased wear on the parts. Drill a gas port off center, and your gas tube will interfere with the carrier key, causing wear and a lack of reliability. Gas blocks are drilled and installed on the barrel to make sure everything will line up perfectly when it arrives at the weapon assembly station or your house.

▲ Noveske Rifleworks' shop mascot keeps a close eye on the blanks, which will later be turned into hyper accurate barrels that the company is known for.

The final step on the barrel is stamping and numbering. This allows barrels to be tracked when customers call in. It also allows Noveske to make sure they are able to detect counterfeit barrels. And, yes, other companies have actually marked their barrels as being from Noveske when they weren't. Thankfully that was a few years ago and it was dealt with, so there's no need to get paranoid about your barrel. The fakes were all accounted for. As John Noveske told us before his untimely death, imitation isn't always flattery — sometimes it's just theft.

Chris Quigley then took us over to final QC, where we met up with two brothers, Jeremiah and Matt. Each station inspects the work that is handed to them before they do their task, but final QC goes over everything yet again, ensuring nothing goes out that doesn't meet Noveske standards. The brothers bring more to the barrel

shop than just a critical eye — they also bring the shop mascot, their dog.

While the barrel shop was clean, the CNC shop was brightly lit and almost surgical in its cleanliness. Here we were able to see the various stations and programs as the Gen III upper and lower receivers were getting machined. We watched as Noveske rails were being machined in all three styles and various lengths; this was the first time we were able to see and handle the newest Noveske Hybrid rail. The CNC shop also tumbles and cleans everything to get it ready for anodizing and laser engraving, which is just a few doors down.

Taking the Gen III platform to the last stage of its manufacturing cycle, we checked out the Cerakote process. There were parts hanging in racks, ready to be sprayed and already "gassed." Items are often preheated before they are sprayed, but gassing goes a step further and actually burns out any impurities that may have been left behind on otherwise clean items before they are coated. With their materials mixed up and test patterns sprayed to verify everything is operating correctly, the applicators made quick work of receivers and rails.

LEGO TIME

Now we were ready for assembly. We were told we could build whatever configurations we wanted to play

▲ In addition to playing with the Noveske Infidel in FDE that's featured in these pages, we got intimate with this 16-inch Gen III in black. Pictured here, Noveske's Chris Quigley fires some rounds downrange.

with. Like kids in a Lego store, we assembled the Gen III Infidel in FDE with the long NSR rail and new KX5. The Infidel is a 13.7-inch barrel, with a KX5 muzzle device pinned and welded to stretch to a legal overall length of 16 inches. The 15-inch NSR rail encapsulates the KX5, leaving just enough of it hanging out the front to look like a cannon. We selected a Noveske stainless steel option for the barrel, but it is our understanding the 13.7-inch barrels will also be available in the chrome-lined machinegun steel, which is a bit thinner in profile.

We chose standard options for most everything else: Magpul ACS stock, Raptor ambi charging handle, Noveske QD end plate, and ALG Defense ALG Combat Trigger (ACT). We went with a Magpul MOE grip just because we like them and don't need the extra pieces the Magpul MIAD offers, though we did note the Gen III Noveske models come with a MIAD grip. We topped it off with the new Magpul MBUS PRO sights.

Next up was a 16-inch Gen III with quad rail. In this case, we were seduced by a 16-inch Noveske RECON barrel that had been Cerakoted black. Instead of the standard SilencerCo mount flash suppressor, we wanted to use the SilencerCo muzzle brake (also can-compatible). We also used a CTR instead of the standard STR stock — after all, variety is the spice of life! For our testing with this one, we would be using a Trijicon Accupoint 1-4x scope and had no interest in iron sights. We were pretty sure we weren't going to be attacked by terrorists during test firing. The Magpul MOE grip went on this build, as well.

RANGE TIME

Range time is always happy time, especially when the guns run like they should. This session was smooth sailing, and there were zero malfunctions of any kind. The ACT trigger broke cleanly, and though it isn't a match trigger, both units felt like super clean USGI-style fire control setups. To get the sad part of our story out of the way, we were *not* able to conduct accuracy testing. Based on the time available, we had to choose between doing a little shooting for accuracy, or doing a lot of shooting to see how the Noveske Gen III lineup functions. We chose the fork in the road that led to high round counts, which seemed legit as Noveske is already known for highly accurate barrels.

Of the two weapons, the higher round count went to the FDE Infidel, which came in at 970 rounds total, with the black carbine being fired 540 times. The ammunition used for this testing was a combination of PMC Bronze, Winchester Q3131, Q3131A, Federal T223E 55-grain TRU Tactical, Speer 64-grain Gold Dot, Black Hills 68- and 77-grain (blue box), 5.56 M855, 5.56 M856, and a little M196. All magazines used during this testing were Magpul PMAG, with a mix of old and new generation models.

We noted several things that were no great shock to us. The Noveske Infidel with the new KX5 had very little muzzle flash with some ammunition, but with others like Q3131, the flash was more noticeable. The same can be said for the muzzle brake from

NOVESKE RIFLEWORKS N4 INFIDEL

CALIBER: 5.56mm

MAGAZINE CAPACITY: 30 rounds

BARREL LENGTH: 13.7 inches

OVERALL LENGTH: 35 inches (31¾ inches with buttstock collapsed)

WEIGHT: 7 pounds

MSRP: $2,595 (Approx.)

URL: www.shopnoveske.com

FEATURED ACCESSORIES:
Trijicon ACOG TA47-2
$1,168

PRICE AS FEATURED:
$3,763 (Approx.)

SilencerCo. Concussion and muzzle blast from the KX5 was nicely directed forward, and it quickly became taken for granted until someone used a weapon equipped with a standard flash suppressor or muzzle brake. The SilencerCo brake made for a very flat shooting weapon, with even one of the most novice shooters being able to successfully engage steel plates 50 yards away with rapid fire.

WE'LL TAKE TWO

At the end of our time with the Noveske crew and their weapons, we learned a few things had evolved, but in some ways things hadn't changed at all. The drive, passion, and excellence of the company continue. The weapons themselves have been upgraded in design as well as fit and finish with the introduction of the Gen III line, and tradition is maintained with the continuance of the Gen I line.

However, the most important aspect from our point of view seems to be the increased production capacity, which will allow a much greater amount of product to get out to the end users. How much did we like our Gen III sample guns? We bought them. Both. R

Noveske's Sheri Johnson gets tips from the legend himself, Clint Smith.

ENTER THUNDER RANCH

While visiting Noveske headquarters, we were asked by company general manager Sheri Johnson if we wanted to go visit Clint Smith. For those who don't know, Smith is the director of the Thunder Ranch range complex, a Marine Corps combat veteran, a former SWAT officer, and a well-respected tactical trainer. We didn't need to be asked more than once if we wanted to meet this legend.

Arriving at Thunder Ranch, we were introduced to Clint and his wife Heidi and taken on a tour of their facility.

If you're familiar with Thunder Ranch, you know it moved from Texas to Oregon. The move allowed for an entirely new range complex to be built, and it is one that rivals or exceeds the best of the law enforcement or commercial ranges we have seen.

If you have never been inside a multi-story shoot house, or engaged targets from anything other than a flat range, Smith's Terminator 3 Tactical House will open your eyes. Walking through the ranges and listening to Clint, we realized his passion and experience in the firearms industry stand out for good reason. After a great dinner with a few great people, we went outside and shot a few pictures while Sheri and Clint were using the Noveske Infidel. In hearing how Clint likes the Noveske weapon platform, we weren't shocked at all to find the Thunder Ranch carbine is made for him by Noveske. Naturally, there will be a Gen III version out shortly.

➔ www.thunderranchinc.com

➔ www.shopnoveske.com

KNIG

ENHANCED COMBAT CARBINE

A 16-Inch Precision Semiauto? Don't Mind If I Do...

By Keith Wood
Photography by Iain Harrison

On my 13th birthday, I was fortunate enough to spend the day in the presence of a firearms legend. Eugene Stoner hosted me in his home office and spent several hours with me discussing and demonstrating various elements of firearm design, as well as the advantages and disadvantages of many common rifles and handguns. Besides the many rare firearms he allowed me to fondle (including a least three prototypes), I remember with great detail the rifle that he literally had on the drawing board that day in 1990. He was co-developing it with a manufacturer a few miles up the road in Vero Beach, Florida, and spent much time explaining the advantages of the rifle and its accessory rail system. That rifle was the SR-25, and the manufacturer was C. Reed Knight Jr. of Knight's Armament Company (KAC).

Even as a kid, I was able to understand the value of a 7.62mm precision semiautomatic rifle with the ability to mount a myriad of accessories via a modular rail system. Fast-forward more than 20 years and KAC has taken the SR-25 concept to a whole new level of versatility, while maintaining the core elements of the Stoner design. The SR-25 is an evolution of the 7.62mm AR-10 that actually predates the AR-15/M16. The SR-25 deleted the trademark carry handle, making the platform far more optic-friendly and featured various design enhancements that increased accuracy. The original SR-25 was widely used by special-operations units and law-enforcement tactical teams — early adopters included the U.S. Navy SEALs, who used the rifle in over-watch roles in Somalia. The rifle has since seen widespread military and law-enforcement use, including the Army's M110 Semiauto Sniper System — an SR-25 variant produced by KAC.

HIT'S

As war often does, the global war on terror has forced rapid weapons evolution; in this case, from highly specialized weapons systems into more versatile tools. In the 1980s and '90s mindset, counterterrorist operators stacking for a building entry wielded sub-machineguns such as the MP5, and the snipers covering their movements often used long and heavy bolt-action rifles. Experiences in Iraq and Afghanistan established that such specialization didn't work so well when troops had to patrol through complex urban and mountain environments to reach a target building, and a sniper may have to fight his way into or out of a suitable firing position. The sub-gun yielded to the carbine, and the dedicated bolt-action sniper rifle has, in many cases, given way to the precision semiauto.

ECC features all the good stuff - ambidextrous controls, free floated, dimpled 5R barrel, enlarged charging handle and offset irons. Aftermarket improvements needed? Zero.

Offset iron sights fold out of the way when not in use and are adjustable for windage and elevation.

TWISTS AND TURNS

Born of those combat experiences, the SR-25 Enhanced Combat Carbine (ECC) is the latest evolution of Stoner's AR-10 design. The core of the ECC is a 16-inch, medium-taper 4150 ordinance steel barrel with 5R rifling and a 1-in-11-inch twist. The 5R rifling, which was made popular by the Army's M24 sniper rifle, uses sloped polygonal lands to form a tight gas seal and to decrease bullet friction. Essentially, it puts less stress and deformity on the bullet and reduces copper fouling that degrades accuracy — a concept that has been around since Metford polygonal rifling was first employed in England during the 1800s. To keep fouling in check even further, and to ensure reliability in dirty operational environments, the bore and chamber of the ECC are chrome-lined. What makes the ECC's barrel so striking and unique, though, are the ball mill cuts that cover the entire surface. These cuts work like fluting, reducing the weight of the barrel as well as providing increased surface area to dissipate heat.

In concept, this method is superior to longitudinal fluting as it imparts less stress on the barrel steel. I have no empirical basis to substantiate or refute this process, I just know that it cuts weight and looks damn cool. The barrel is topped-off with a QDC flash suppressor that allows for quick attachment of Knight's own line of

sound suppressors. Spring-loaded ball bearings on the inside of the suppressor's locking collar engage into mill ball cuts on the flash hider and lock the can in place with the twist of a ring. The system is fast, secure, and repeatable.

Modularity has been a hallmark of the SR-25, and all KAC designs for that matter, since its inception — the rifle length URX railed fore-end continues that theme. The URX is free-

floated so no bipods, lasers, or other accessories mounted on it will change the barrel harmonics and cause a shift in zero. The uninterrupted top rail provides plenty of real estate for mounting the user's choice of day and night optics. The 3, 6, and 9 o'clock positions of the fore-end feature Picatinny rail sections fore and aft, where accessories are likely to be mounted, but the fore-end narrows into slick aluminum slabs to cut weight and to provide a diameter that is easy to grip with support hands of all sizes. The rail features several integral mounts for QD sling swivels, and is drilled and tapped for the addition of grip panels or even additional rail sections on the smooth surfaces.

HANDS ON

Our test rifle came equipped with KAC's 45-degree offset micro folding sights, mounted at the 1:30 position. If the rifle's primary optic fails or has too much magnification for the task at hand, just roll the rifle 45 degrees to the left, flip up the sights, and keep shooting. The sights are very low profile and a thing of beauty, despite being designed for auxiliary use. The aperture rear and square post setup offers an excellent sight picture, not unlike that of the M16A2, and they are fully adjustable for windage and elevation. If you are one of those cursed with being left-handed, you can install the sights on the left side of the weapon, but they will flip forward instead of backward.

The ECC's controls will be familiar to anyone with time behind an AR, though it also features ambidextrous mag release and selector/safety levers. Not only does this make life easy for left-handed shooters, it allows a right-handed shooter who is firing from the weak shoulder to manipulate the controls and reload the carbine without shifting position or exposing himself unnecessarily. The two-stage match trigger breaks at a consistent 4 pounds. I found the trigger to be light enough for long-range precision shooting, but not so light to be a liability in more dynamic scenarios. An oversize bolt stop makes for fast reloads and is easy to manipulate when locking the bolt to the rear.

In keeping with the theme of versatility, we mounted a U.S. Optics 1-8x27mm SR-8c Red Dot scope in a GDI P-ROM L mount. With true 1x magnification and an illuminated reticle, this rifle is capable of fast work at short range and is equally at home making hits well past 500 yards at 8x. With the addition of a bipod from Atlas and a two-point bungee sling from Original Special Operations Equipment, we had an ideal setup for a recce/designated marksman role. Loaded up with a few hundred rounds each of ASYM Precision 178-grain OTM and Federal XM762D 150-grain FMJ, we headed to the range. Thanks to a rifle-length gas tube and a longer-than-standard buffer, the ECC was softer-shooting than I expected. This made for fast follow-up shots at short range and the ability to rapidly reacquire targets at longer distances. I've had nothing but positive experiences with AYSM's Match ammo in the past, so I used it for all of the accuracy evaluations and long-range shooting. The ECC was a solid shooter, with five-shot groups consistently at or under an MOA. With the Atlas bipod loaded with some forward pressure, the EEC was dead steady from prone, and 400-yard hits on MGM steel targets were fairly automatic. Though I wasn't able to shoot the ECC past 500 yards, in my experience, the 8x optic and 7.62mm ballistics out of

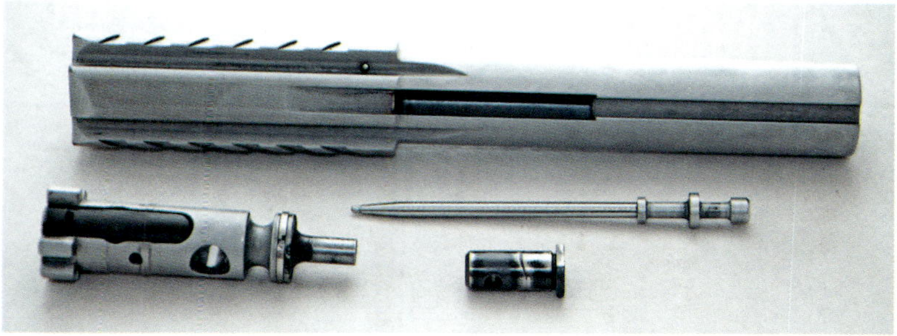

The AR10 platform has always struggled to delay bolt unlocking long enough for the bullet to exit the bore. KAC uses a 4.5-ounce buffer and adds considerable mass to the ECC bolt carrier to increase inertia.

the 16-inch barrel would likely make this setup very effective out to about 800 yards. With more powerful optics (or a talented shooter), that envelope could surely be pushed a bit further. As for reliability, the ECC functioned with 100-percent reliability during all aspects of our testing.

CALIBER CONSIDERATIONS

The ECC's best attribute is its size and weight relative to its capabilities. When I asked SEAL sniper instructor and Centurion Arms owner Chief Monty LeClair his opinion on this class of rifles, he put it to me this way, "On the modern battlefield, overspecialization is a no-go. A guy may have to fight his way through a house to get to an over-watch position, and that's tough with a big rifle. Something fairly compact with the ability to make precision shots is a huge benefit." With that in mind, I ran the ECC through some short-range drills and it performed admirably. A 9-pound rifle doesn't swing as fast as a 10.5-inch SBR, but the ECC handles reasonably well in tight quarters.

The 5.56mm rifles and carbines dominate the U.S. and western military/law enforcement marketplace for good reason, but the cartridge does have its limits. The 7.62x51mm hits harder, penetrates cover more effectively, and is capable of maintaining accuracy at greater distances than its NATO successor. A 7.62mm carbine that can be carried as a primary weapon, but also employed in a sniper role would be a valuable asset to any small unit. The ECC fits that role perfectly. For a civilian looking for one gun to do it all, the ECC meets the task equally as well.

Like all of Knight's products, the ECC is a well-designed, solidly built piece of kit. Like all of Knight's products, the ECC is also expensive. But this is a carbine that can be carried for extended periods of time and brought to bear effectively at short and long ranges alike. You can buy a lighter rifle, you can buy a more accurate rifle, and you can certainly buy a cheaper rifle — but you'd be hard-pressed to find a rifle with this combination of accuracy, mobility, durability, and adaptability. R

KNIGHT'S ARMAMENT COMPANY SR-25 ENHANCED COMBAT CARBINE

CALIBER: 7.62x51mm
MAGAZINE CAPACITY: 20 rounds
BARREL LENGTH: 16 inches
OVERALL LENGTH: 35.75 inches
WEIGHT: 9.12 pounds
MSRP: $5,750
URL: www.knightarmco.com

FEATURED ACCESSORIES:
U.S. Optics 1-8x27mm SR-8c Scope - $2,495
GDI P-ROM L Mount - $565
B&T Industries Atlas Throw-Lever Mount Bipod - $280
Knight's Armament 45-Degree Offset Sights - $281
Original Special Operations Equipment Two-Point Bungee Sling - $50

PRICE AS FEATURED: $9,421

ADM UNIVERSAL IMPROVED CARBINE

By Freddy "Sue" Osuna
Photography by Kenda Lenseigne

A Carbine for the Combat Tracker in All of Us

EDITOR'S NOTE: We covered the essentials of combat tracking in RECOIL Issue 9, where instructor Freddy Osuna explained how to examine tracks in order to hunt down human prey. Combat trackers typically operate at the very tip of the spear, ahead of their supporting troops while following an elusive enemy across broken country. As such, the requirement for a lightweight, effective weapon is paramount, so when American Defense Manufacturing (ADM) announced their 6-pound carbine, we figured we knew just the guy to try it out.

A combat tracker's primary weapon requirement is for a lightweight weapon with both the ability to quickly neutralize an immediate threat and the capability for accurate, midrange target suppression — to either kill the mutha, break contact, or lay down fire until the cavalry arrives. The American Defense Manufacturing (ADM) Universal Improved Carbine fits the bill.

After attracting attention at SHOT 2014, ADM's new 5.56mm carbine caught the eye of RECOIL contributor Aaron Cowan, and we profiled a preproduction version of the rifle on RECOILweb.com. Now that it's ready for primetime, we invited Adam Pini, vice president of sales at American Defense Manufacturing, out to Cowtown range to join us for a day of jaw jacking and mag slapping. Put your feet up for a synopsis of the features and our experience shooting the rifle that's call signed the ADM Universal Improved Carbine MOD 2.

ADM was warned that their progeny might end up as the cover gun for this issue, so they stepped up and sent it to their Cerakote booth for the full-house MultiCam treatment. Every part was masked and sprayed, including a B5 SOPMOD stock, Trijicon VCOG 1-6x24 scope, Griffin Armament Fail Safe sights, ADM Low Light rail, Griffin Armament M-4 SP Suppressor, and Surefire Scout light. The contours of the rifle along with its accessory placement and the camo job just screamed, "Pick me up!" So after probing for pressure release devices and determining said object was safe to grab, we took the rifle into custody.

FULL FEATURED

In a world of plain vanilla ARs, the UIC offers a dash of habanero. Spicing up this offering is its fully ambidextrous control suite, adding an extra level of piquancy to the mix. The ambidextrous safety, conveniently and easily swept with the inside edge of the firing index finger, is nice to have as an alternative option, particularly when it comes to putting the weapon on safe when working around teammates. In this case, more is better, so it's good to see that all hand sizes are accommodated with several different options of widths, lengths, and heights for the safety lever.

It sweeps from safe to semi at an ergonomic 60-degree angle as well, rather than the typical 90. The magazine well is beveled at 20 degrees by request and insight from competi-

MULTICAM
MULTICAM

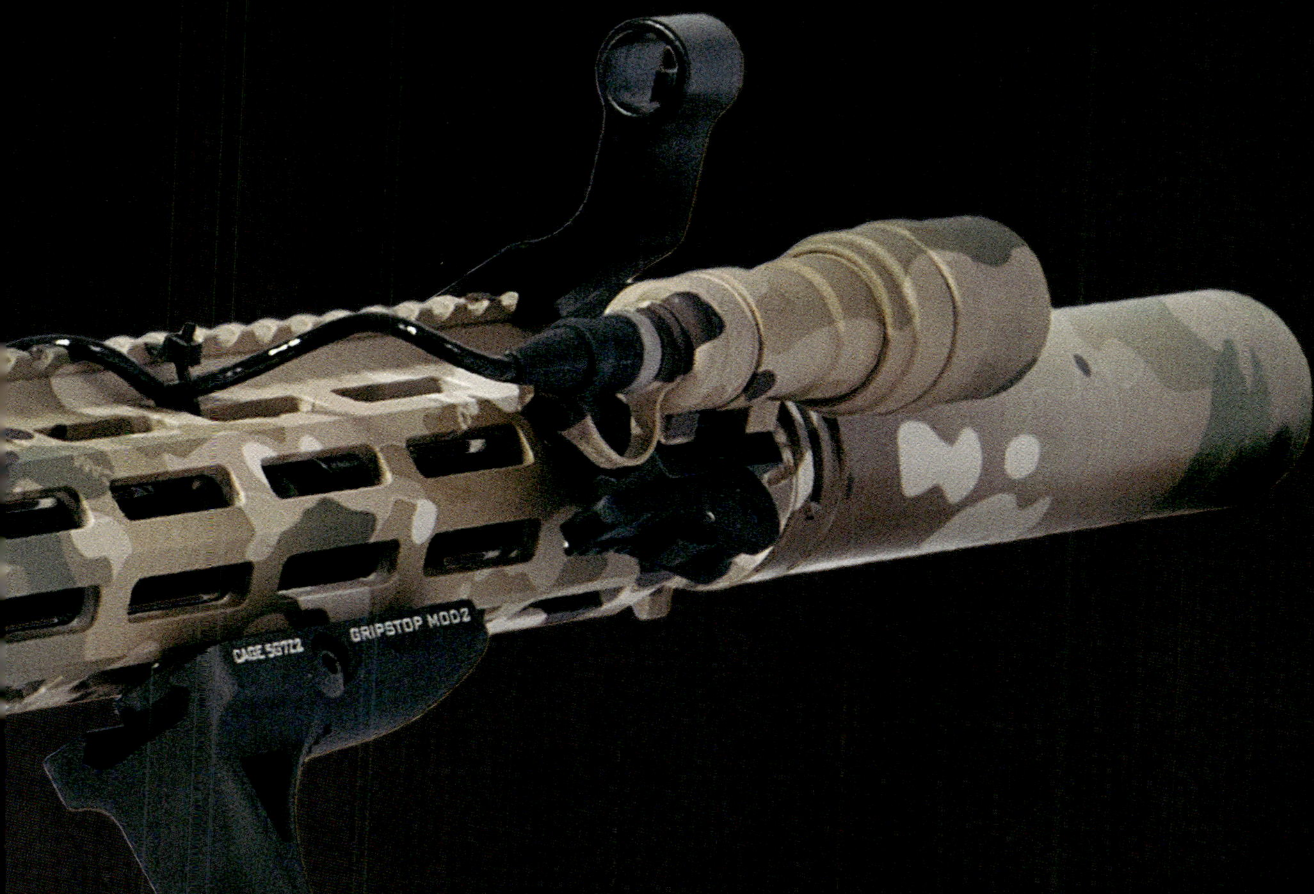

Note the ambidextrous bolt catch lever behind the magazine release. It provides more functionality than add-ons like the Magpul BAD lever in a cleaner, completely integrated fashion. ADM hit a home run with this feature.

tive shooters, and of course it sports ambidextrous magazine releases. The ambi theme is carried over to the charging handle, which has no manual latch, instead being secured by a "force to overcome" detent that simplifies manufacturing and results in fewer stressed parts. After a brief tour of the exterior, the most prominent feature turned out to be an extra slot cut into the fence above the mag release. It housed the ambidextrous bolt catch — and of all the ways to make a true ambi AR, this is the best we've encountered so far. It is slicker than snot on a doorknob. While firing off hand, we deliberately blocked the ejection port with the support hand and interrupted the ejection of a casing. With the rifle pointed in a safe direction, we instinctively slapped the magazine, observed the malfunction, pulled and locked the charging handle to the rear (with the non-firing hand on the bolt catch), removed the source of ammunition, cleared the obstruction, inserted a fresh magazine, slapped the bolt release (again with the non-firing hand), reacquired the sights, and successfully fired. No biggy there, as years of muscle memory took over.

Pini immediately suggested that with the ambidextrous controls, we could use the firing side hand to both lock and release the bolt, thus placing the rifle into battery much sooner as that hand would never leave a firing grip. We played around with the new technique, and frankly, it's a no-brainer. Anything that keeps that shooting hand where it belongs — i.e. hand on the grip, head over the gun, and gun in shoulder — is going to gain time in a firefight. A second is an eternity in combat. This one feature alone is the most appealing function of this rifle. If this were my primary weapon, I would gladly invest in the thousands of rounds it would take to ingrain the remedial actions upon a type 2 or 3 malfunction, as it's just that much faster and more positive.

On the range, weight and feel were as advertised — the gun felt like it was milled out of one piece of metal and had a nicely neutral balance. When a lightweight buttstock is paired with a longer or heavier barrel, it often produces a muzzle-heavy gun that is only exacerbated when the user adds accessories. This becomes tiring to carry and is slower to get on target. ADM used a good-quality Wilson stainless barrel blank that they contour to a lightweight profile, producing a quick-handing carbine that still provides good accuracy, even as it heats up. The selection of quality parts extends to the bolt carrier group, made from ASTM 9310 alloy and subjected to a QPQ nitride treatment.

HOME ON THE RANGE

After a pre-fire inspection and determining there was no battle rattle

due to loose hardware, we conducted a function check and took the UIC out to zero at 100 yards with Geco 55-grain ammo. There seemed to be something special about my fellow Marine's paper target on the lane next to mine, because he was insistent about my not shooting it — or else. So naturally I shot it, triggering a burst of mumbled cursing that seemed to last quite a while

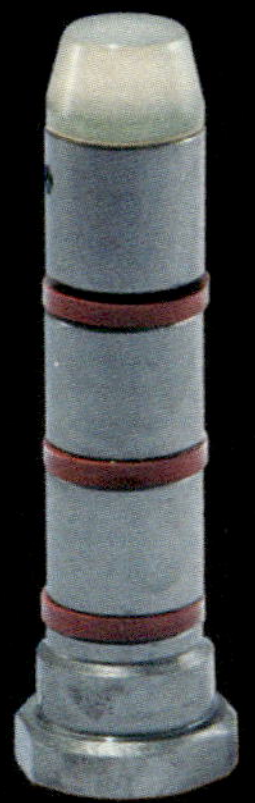

until someone could distract him with something shiny. After 15 rounds, the rifle was holding one MOA as advertised. So we then switched to hammering on a 6-inch steel popper.

Following the pleasant introduction, we meandered to the top of a hill that overlooked a 1,000-yard known distance range, where 10-inch gongs were placed within, on top, and on the sides of an arroyo. Environmental conditions were ideal, with 5 to 15 mph half value (from 7 to 1 o'clock) gusts in the morning and occasional 10 to 25 mph swirls in the early afternoon. Targets were all highly visible, with white spray paint giving immediate feedback on impact.

Upon cresting the hill with the UIC, there were three immediate threat targets in our field of view:

AMERICAN DEFENSE MANUFACTURING
UIC MOD II

CALIBER: 5.56mm

MAGAZINE CAPACITY: 30 rounds

BARREL LENGTH: 12.7 inches (16-inch barrel available)

OVERALL LENGTH: 30 inches (stock collapsed)

WEIGHT: 6.2 pounds

MSRP: $1,850 (Black), $2,000 (Cerakote colors)

URL: www.americandefensemanufacturing.com

FEATURED ACCESSORIES:
Trijicon VCOG 1-6x24 Riflescope - $2,800
ADM AD-B3 Modular Base Scope Mount - $80
Griffin Armament M4SD II Silencer - $995
Griffin Armament Fail Safe Angle Sights - $195
Surefire M300 Mini Scout Light - $369
ADM AD-SF SureFire M600B/M300A Mount - $67
Lanco Tactical Grip Stop Mod 2 - $65

PRICE AS FEATURED: $6,571

one to the direct front at 12 o'clock, 100 yards across the arroyo at eye level; the next at 4 o'clock, 200 yards in the shadow of a Palo Verde tree; and the last target at 2 o'clock at 300 yards on the far side military crest. We engaged the closest target with three rounds off hand with three hits, swept to the second at the right by dropping to a knee and placed two out of three hits, then hit the deck in prone and engaged target number three for two out of three hits. Pretty much what life in my MOS consists of….

The most notable impression following these target engagements was the rapid sight recovery — getting quickly back on target after each recoil impulse, followed closely behind the reset of the ADM two-stage, 4.5-pound trigger. Unlike most carbine-length gas ARs, this carbine is smooth. How smooth? Think Scarlett Johansson's inner thigh smooth. Apparently, this is due in part to an improved buffer, which can be tuned to the owner's preference by adding or removing weights. According to Pini, "the ADM reliability-enhancing buffer provides non-reciprocal weight matching the rifle length buffer, coupled with a platter of weights that impact in a staggered sequence to eliminate bolt bounce without resorting to springs or hydraulic systems, which can wear and fail." From introduction to application, we were satisfied with this rifle's performance and in our brief experience with the UIC can confidently attest to the quality of its build. At a price point of around $1,850, it's not cheap, but you're getting a lot of rifle for your dollar. Should this be outside of the weekly budget, you could be the lucky patriot who is going to win this very rifle, as ADM intends to raffle it off during SHOT 2015 with all proceeds going to charity. Check their Facebook page for updates. R

ABOUT THE AUTHOR

Freddy Osuna is a USMC combat veteran, scout sniper, outdoorsman, and author. He is a respected authority on the subject of combat tactical tracking and provides training to military, law enforcement, and outdoorsmen alike through his school, Greenside Training based in Benson, Arizona.

HOT WASH

The American Infantry Rifle and 25 Years of Lessons Learned

By Tom Marshall
Photos by Iain Harrison

About the time this issue hits the newsstands, the U.S. Special Operations community will likely be taking a look back at one of the most high-profile operations in their history: Operation Gothic Serpent, which included the infamous Battle of the Black Sea, made famous by the book-slash-movie Black Hawk Down. That mission, which took place in October of 1993, is officially 25 years old this fall.

This carbine was considered state-of-the-art around the time Meatloaf topped radio charts with "I'd Do Anything For Love (But I Won't Do That)." If that doesn't make you feel old ...

Several veterans of that operation are currently active in the firearms industry and have given their historical accounts of the mission to various media outlets. Instead of trying to retell someone else's war story, we wanted to take this anniversary to examine the progress of America's everyman rifle over the ensuing two-and-a-half decades, and perhaps reflect on just how good we have it now.

BLAST FROM THE PAST

As the rise of the retro rifle continues to gain momentum, several companies are now producing period-themed AR-pattern rifles to commemorate past iterations of Stoner's most famous design. Troy Industries was one of the first to offer an out-of-the-box solution to collectors and enthusiasts wanting a "period" rifle with their My Service Rifle line, commemorating famous military operations, and the associated rifles used to win the day.

Their recent release of the M16A2 SFOD-D carbine made an all-too-appropriate cornerstone for this project. This no-frills rifle was state of the art at the time it was used by small-team elements of the U.S. Army and Air Force in the late '80s and early '90s. It's a 14.5-inch barrel, carbine-length gas system affair with traditional CAR handguards, iron sights, and an A2 carry handle upper. The gun ships with a length of rail mounted on both the carry handle and the 6 o'clock position at the forward end of the handguard.

As a preface to all of you firearm historians out there, please note that this was an "in the spirit of" build and does features accessories in the style of this period, as opposed to the actual items. Attempting to procure the actual lights, sights, and mounts from two-plus decades ago was hardly conducive to deadlines or production budgets. So, in several cases, we had to make do with "close enough." Good enough, as the saying goes, for government work. This particular Gothic Serpent sample is outfitted with a SureFire 6P complete with a whopping 60-lumen incandescent bulb, mounted on a single scope ring with their push-button tactical tail cap. The optic is an Aimpoint 9000, which uses the longer tube style of the older 5000 with updated electronics.

While the idea of mounting a light to a weapon isn't exactly new, the technology to do so in a manner that's both convenient and ergonomic is a relatively recent development. As late as the early years of Operation Iraqi Freedom, line units were using duct tape and hose clamps to hold D-cell mag lights onto their rifles. The SOF community, having a larger budget and more time dedicated to R&D, found that you could use weaver scope rings to mount the then-new smaller lights made by SureFire onto their guns. Certainly better than the methods used by conventional units even a decade later, this small measure of convenience came with two primary pitfalls — actuating the light and lumens.

Though night vision, and the earlier starlight technology, dates back to Vietnam and somewhat before, dedicated night-fighting gear isn't a catchall for "intermediate" lighting situations. Think about entering a dark room in the middle of a bright desert afternoon in Africa. You need some kind of artificial light to see your target, but early night vision goggles — prone to washout or permanent damage from ambient light through a window or hole in the ceiling — were the wrong answer. So weapon lights became the best compromise.

Even though any advantage is better than no advantage, less than 100 lumens doesn't buy you much reaction time. As your eyes are rapidly adjusting from bright light, to no light, to a little bit of light the "increased" ability to identify friend from foe is marginal at best. Tape switches were available at the time, but far from universal and far from reliable. They had to be taped on and, if you've ever had a piece of tape peel off something in the heat, you know that taping things together isn't the most ironclad attachment method.

Once you get the light mounted, you have to be able to actually turn it on. With the light at the bottom of the handguard, thumb activation is out of the question. To make this placement work, we had to shift our support handgrip to just past the magwell and use the index knuckle of that hand to trip the light. It works, but not well. While firing, we had trouble keeping enough pressure on the switch to keep it on. The other option is to twist the tailcap for constant-on, but then you run into the fairly obvious issues of battery life, and of giving away your position between engagements.

Once you can see your target, you gotta hit it. The early electro-optical sights, also of Vietnam vintage, were a huge boon for rapid shots under tight constraints. The optics themselves, to include the Aimpoint 3000s and 5000s of the *Black Hawk Down* era, didn't have the kind of battery life or reliability that we now expect from any red dot worth its salt. But mounting them on an A2-style receiver created an additional issue: height over bore.

For the uninitiated, height over bore is exactly what it sounds like. Mounting your scope several inches above your barrel creates the need for both mechanical offset when you zero as well as for manual holdover when trying to make precise shots — like headshots, which are a common point of training for hostage rescue units. Furthermore, these high-mounted optics require a "chin weld" on the stock, which is unnatural, uncomfortable, and offers a floating sight picture at best, particularly while shooting on the move.

LATEST AND GREATEST

As a demonstration of the technical progress that's been made in configuring the AR or M4-style rifle, we contrast Troy's My Service Rifle SFOD-D gun to their own cutting-edge carbine, the SOC-C. The SOCC (Special Operations Compatible Carbine) also sports a 14.5-inch barrel chambered in 5.56mm — which is squarely where the similarities end. The SOC-C features a mid-length gas system. Recent testing by USSOCOM has proven what the commercial market has known for years —that the longer gas tube makes for a cleaner and softer shooting weapon.

The SOCC covers that gas tube with a 12-inch M-LOK handguard. This single feature offers the warfighter a level of modularity that hasn't been known since the M16's

Synergistic advances in handguards, lights, and forward grips provide a support-hand hold that's more ergonomic and offers better control over the weapon.

Things like lower height-over-bore and shorter overall length give the SOCC carbine a distinct edge over it's partner. Internals and fire controls are also highly improved over Mil-spec.

introduction six decades ago. Now you can mount your lights and any other accessory wherever you want. In our case, we used SureFire's new 600DF weaponlight attached to the rifle by way of an Arisaka Defense inline mount. The 600DF produces 1,500 lumens, which not only restores small rooms to broad daylight conditions at the push of a button, but can probably be used to signal low-flying aircraft or heat up your MRE.

When Super 6-4 went down near the Bakara Market in Mogadishu, soldiers had to mount a rail to the handguard, a scope ring to the rail, and the light into the scope ring. This system creates poor ergonomics and multiple points of failure for your light to shoot loose or fall of completely. With the 600DF/Arisaka combo, the mount is screwed directly into the body of the flashlight, and then attached directly to the handguard. Not only is this a simpler system less prone to mechanical failure, but the advent of modular handguards provides adjustability in where the light is places, both lengthwise along the fore-end and around its circumference. The biggest single benefit to come from this advancement is that, now, you can configure the gun around the operator's natural stance and hand placement instead of changing how you fight just to accommodate a flashlight.

Optics have gotten smaller, smarter, tougher, and more diverse in the last 25 years. Our SOCC sports an Aimpoint Comp M5. It's their smallest and most efficient rifle-mounted red dot. With battery life measured in years and a slew of brightness settings that include night vision compatibility. The move from carry-handle upper receivers to full-length top rails provide a laundry list of benefits on a fighting rifle. The aforementioned height-over-bore issue all but disappears. This simplifies zero. It also simplifies unconventional shooting positions like shooting over or under a barricade and allows a proper cheek weld. Additionally, the full-length top rail allows end users to utilize different types of optics. The vast increase in mounting space means that force multipliers like variable-power glass and clip-on thermal or night-vision units can be mounted quickly and securely with no tools, as the mission changes.

ALL THE SMALL THINGS

While lights and sights were our two most obvious observations, there are other less prominent improvements that are equally important. One is the advent of ambidextrous controls. While, statistically, the number of left-handed shooters is pretty low throughout the ranks, if you happen to have one on your team you want them to reap all the same benefits everyone else in the stack does. Ambi selector levers, charging handles, and mag and bolt releases all create a perfectly mirrored manual of arms, regardless of which hand is pressing the trigger. But it's not only southpaws who get something out of it.

The advent of urban warfare has forced U.S. soldiers to enter a battle space full of walls, windows, and hard angles. Being able to transition your carbine from strong side to support side as you adapt to available cover offers a very real increase in soldier survivability. Ambidextrous buttons and switches allow all shooters to switch-hit off of barriers without having to change anything about how they drive their gun.

The last, but perhaps most critical upgrades we'll discuss come in the form of the almighty bang switch. After executing proper stance/grip/sight alignment/sight picture, trigger press is the shooter's last physical input into the weapon before that round leaves the barrel. Sloppy or harsh trigger press can throw a shot even if you do everything else right. This becomes a literal matter of life and death for units that fight in very close quarters where hostages and innocents are all in play.

The M16A2 SFOD-D sports a standard Mil-spec trigger that was delightfully rocky and inconsistent. By comparison, the SOCC comes out of the box with a Geissele G2S trigger. While not marketed as a match trigger per se, it offers a gliding smooth take-up with a consistent break that snaps like a carrot each and every time. It's this consistency and predictability that gives a shooter an opportunity to improve their marksmanship more quickly, as well as imparting a confidence that the trigger will do exactly what you want it to every single time — a not insignificant comfort when entering situations measured in tenths of a second.

Newer shooters, and older ones who have embraced progress, get quickly adjusted to the ease with which a modern, properly configured rifle can be run hard under demanding conditions. While the events of Operation Gothic Serpent can be labeled as both tragic and heroic, the lessons learned from those units and their experience cobbling together a "best possible" solution with the parts they had set in motion a ripple effect that helped birth the cutting-edge carbines we now use to defend our country and our homes. R

TROY SOCC CARBINE 14.5"

CALIBER: 5.56mm
MAGAZINE CAPACITY: 30 rounds
BARREL LENGTH: 14.5 inches (16 inches with pinned Troy flash hider)
OVERALL LENGTH: 34.5 inches (stock extended)
WEIGHT: 6.75 pounds
MSRP: $1,599
URL: worldoftroy.com

FEATURED ACCESSORIES:
Aimpoint Comp M5 - $1,079
Surefire 600DF - $299
Cain Arms Mini Vertical Grip - $15

PRICE AS FEATURED: $2,992

TROY M16A2 SFOD-D

CALIBER: 5.56mm
MAGAZINE CAPACITY: 30 rounds
BARREL LENGTH: 14.9 inches (16 inches with pinned A2 flash hider)
OVERALL LENGTH: 36 inches (stock extended)
WEIGHT: 6.1 pounds
MSRP: $1,299
URL: worldoftroy.com

FEATURED ACCESSORIES:
SureFire 6P - N/A
Aimpoint 9000 - $503

PRICE AS FEATURED: $1,802

Even decked out with a camera and two optics, the V7 LW still comes in under 8 pounds.

LIGHTWEIGHT VERSATILITY

Daniel Defense Do-It-All V7 LW

By Recoil Staff

Daniel Defense's M4 Carbine V7 LW is built around the premise of keeping it light in weight.

From a distance, all AR-15s look the same: a mostly uniform amalgam of aluminum, steel, and plastic, all dressed in black. They're all manufactured to the same set of specifications, and are comprised of the same components: a barrel, receivers, a grip and a stock. With so many competitors producing a similar product, how is it that Daniel Defense stands out in such a crowded market? To excel in a domain in which your product looks very much like your competitors', the company chose to concentrate on one important factor: build quality.

Whether you are a casual shooter or make your living in law enforcement, having a reliable carbine is essential. Daniel Defense, known for high-quality products at reasonable prices, supremely delivers in that department. Initially famous for its high-quality, railed handguards, the Black Creek, Georgia-based company has quickly grown to become a successful commercial rifle and accessories manufacturer as well as a supplier to the U.S. and UK armed forces. Specializing in AR-15s, Daniel Defense currently offers 17 unique rifles. The M4 Carbine, V7 LW (Lightweight), is one of its newest models.

ATTENTION TO DETAIL

Unlike most of its competition, Daniel Defense is one of the few companies in the United States that has the capability and expertise to actually make its own cold-forged hammer barrels. While others purchase from outside manufacturers, Daniel Defense produces barrels in-house from scratch. Starting with raw steel bar stock, the company hammers its own barrels, producing some of the best in the business. Hammer-forged barrels last longer and have a smoother bore surface than barrels produced through other processes.

1. Daniel Defense's rifle features a satin-like finish.

2. The Castle Nut is staked to keep it from coming loose.

3. Trijicon's TA31H-G ACOG was mounted in the traditional position on the top of the upper receiver.

4. EOTech's MRDS mounted in Daniel Defense's 1 O'clock mount, placing it at roughly 45 degrees from the shooters perspective.

5. To help keep muzzle rise to a minimum we installed a PWS FSC556.

6. Spiked tips of the Noveske K9 Barricade Support.

7. We added to the front end, a Troy Folding Sight and Contour GPS Camera.

The heart of the V7 LW is a 16-inch, chrome-lined, cold-hammer-forged barrel coated in a heavy phosphate finish. The barrel is made of 4150 chrome moly vanadium steel; because Daniel Defense doesn't skimp on quality, their barrels are high pressure and magnetic particle inspection tested. The V7 LW barrel features a midlength gas system capped with a pinned, low-profile gas block. The pinned gas block ensures that it is solidly in place and will not become misaligned under duress. Its one-in-seven twist rate is excellent for stabilizing bullets in the 55- to 77-plus-grain range. Another of the barrel's highlights is its lightweight profile. We measured the diameter of the barrel at the muzzle end at 0.575 inches with a 0.625-inch, gas block. This helps make for a very light carbine.

Daniel Defense's 12-inch, Modular Float Rail (MFR) handguard comes as standard equipment. The aluminum handguard is of the free-float-tube variety, and includes an uninterrupted, Picatinny rail that runs across the entire top of the handguard. Also included are optional, three-inch, Picatinny rail sections that can be mounted by using button-head hex screws backed by specially threaded washers that fit in the slots that dot the entire handguard. The handguard can also be run without the rail sections for an ultralight and snag-free setup.

The barrel and MFR find their home mounted to a nicely finished, flat-top, upper receiver that is machined with M4 feed ramps. Its Picatinny rails are marked with numbers for repeatable placements of quick-detach (QD) optics and other accessories. The upper receiver also houses a forged charging handle and a "DD"-marked bolt-carrier group. Because it is built to last, the bolt is MP tested and the gas key on the carrier is staked.

The lower receiver is fitted with a no-frills, single-stage, USGI trigger that is pretty standard in basic AR builds and not all that remarkable. We've shot with our fair share of USGI triggers and have to report that the feel of this one was on the chalkier side. The pull ranges at about six to eight pounds with a definite reset that can be clearly felt and heard. A Magpul MOE polymer trigger guard comes standard. A very nice touch is the QD swivel-attachment point built into the plate at the rear of the receiver. A standard A2 grip rounds out the lower receiver. The receiver is clearly produced and assembled with care using quality parts and finishing. However, we wished there were ambidextrous controls offered.

The buttstock assembly features a Mil-Spec-sized six-position buffer tube, a heavyweight carbine buffer, and a Magpul MOE stock. The castle nut is staked, which we don't see enough manufacturers do. [Attention to detail is everything at Daniel Defense, and the company excels at it.] At times, a castle nut can come loose, and when it does, your shooting day is over, unless you have the right tools and parts with you. Having a staked castle nut prevents that from happening.

ADD-ON GEAR

The lightweight disposition of the V7 LW lends it well to shooters that are looking for a full-sized carbine that they can shoot all day and not feel the tiring effects of lugging around dead weight. At six pounds, three ounces unladen, Daniel Defense really meant LW when they named this rifle. What the weight savings means to us is that we can outfit the carbine with gear and not feel the effects we would starting out with a heavier rifle.

For optics and sights, we went with a Trijicon TA31H-G ACOG 4x-magnified sight in a LaRue Tactical LT100 QD mount; we mounted EOTech's new MRDS sight in a Daniel Defense 1 O'clock offset rail mount, which puts the MRDS right over the bore for an easy transition to red dot sighting.

LIGHTWEIGHT BARREL TEMPERATURE TEST

Temperature results of 90 rounds fired in rapid succession through the Daniel Defense Lightweight profile barrel.

AMBIENT AIR TEMPERATURE: 60°F

COLD BARREL: 64.8°F

MAGAZINE 1 (30 Rounds): 142.2°F

MAGAZINE 2 (30 Rounds): 212.3°F

MAGAZINE 3 (30 Rounds): 297.7°F

Time for the barrel to return to "cold" (under 70°F): 22 minutes

Three temperatures were taken per listed reading, average temperature shown.

Measuring device: **Mastercool Inc. Infrared Thermometer with Laser**

Daniel Defense offers the V7 LW with or without backup iron sights (BUIS) direct from the factory. We would have chosen the option with the sights, but Daniel Defense only offers fixed BUIS, not folding ones. Knowing that we were going to outfit the carbine with an ACOG that would not clear the height of the rear fixed sight, we opted to go without the factory BUIS and added a set of Troy Industries BattleSight folding sights instead.

Daniel Defense is now in partnership with Contour, a hands-free, video-camera manufacturer, to bring onboard video equipment right to your rifle. We mounted a ContourGPS camera on one of the Picatinny rail sections that is included with the V7 LW. The camera allows a marksman to record his or her shooting for later review. It is especially helpful for training purposes, or you can use it just for fun. The ContourGPS captures not only 135-degree wide-angle HD video, but records your location, speed, and elevation as well.

To keep the muzzle down under rapid fire, we installed a Primary Weapons Systems FSC556 that is an excellent flash-suppressing compensator. The V7 LW comes with a factory A2, birdcage-style flash hider, but we just can't resist our aftermarket goodies. Also mounted on the front end is a Noveske K9 barricade support. Its spiked face lets the user jam it against a barricade for stable, one-point-supported shooting. The K9 was developed by the Noveske Shooting Team for 3-Gun competitors, but we feel that it has applications in casual and tactical shooting as well.

Even after we added on all of these accessories, the unloaded V7 LW still weighed in at only 7 pounds, 11 ounces. That's pretty incredible, considering that it's carrying two optics with mounts, a set of BUIS, an onboard camera, a compensator, and a barricade support. While being a positive factor in the V7 LW's light weight, we did notice that the MOE stock did make the rifle feel a bit unbalanced (front heavy) when the accessories were attached. A slightly heavier stock would balance it out nicely, while still keeping the rifle well under nine pounds kitted out.

AGILE FIRING

At the firing range, we loaded the carbine with various magazines and ammunition to see how it performed. We fed it Federal XM193 55-grain FMJ, 5.56x45mm rounds as well as PMC .223 Remington 55-grain FMJ rounds. The source magazines were a SureFire 60-round, a Magpul PMAG, and a standard-aluminum USGI. The carbine ran all combinations without a hitch. The magazines were easily fed into the V7 LW's enhanced flared Magwell and locked in sure and fast. Both the .223 and hotter 5.56 rounds fed and ejected well. The rifle was nimble to point, and having it in hand or slung over our shoulder for a couple of hours didn't pose a problem. The midlength gas system, coupled with the PWS FSC556 we installed, kept the rifle shooting nice and flat.

Daniel Defense lower receivers (above right) feature an enhanced flared magwell for easier magazine changes.

Trijicon's TA31H-G ACOG allowed us to take precise shots at distance.

Some detractors of lightweight barrels say that lightweight profiles allow the barrel to heat up too fast, contributing to a loss of accuracy. We decided to do a pretty extreme test to see if this was true. We started with the barrel cold and shot four, 5-round groups from 50 yards, then executed three, 30-round-mag dumps, immediately followed by four more, 5-round groups. All were shot with Federal .223, 69-grain Sierra MatchKing BTHP bullets. The first set of 5-round shots from a cold barrel yielded sub-MOA groupings averaging 0.83 inches. Our Rambo-like rampage of 90 rounds of rapid fire heated up the barrel enough to make it clearly smoke and emit a distinct, burning aroma. While it smoked; we initiated our second set of groupings. Looking through the Trijicon ACOG, we saw a heat mirage right above the barrel similar to what you see on hot asphalt on a blistering day. Our four, 5-shot groups with the barrel hot still maintained sub-MOA groups with an average of 0.85 inches, but the overall point of impact dropped an average of one inch. We know what you're thinking: Who shoots groups right after dumping 90 rounds from a burning hot rifle? We did so you don't have to. While the groups hardly changed, the one-inch drop at 50 yards is significant. For those who would rather have less shot deviation due to a heated barrel or for those musclemen who don't care about keeping their rifle lightweight, Daniel Defense also produces a standard V7 with a heavier, government-profile barrel.

After bench shooting the V7 LW, we decided to set up a course of fire with both close-, medium-, and long-range targets. We placed cardboard silhouettes at five, 15, and 30 yards, one 10-inch steel gong at 100 yards, two 18-inch steel squares at 230 yards, and, finally, two additional cardboard

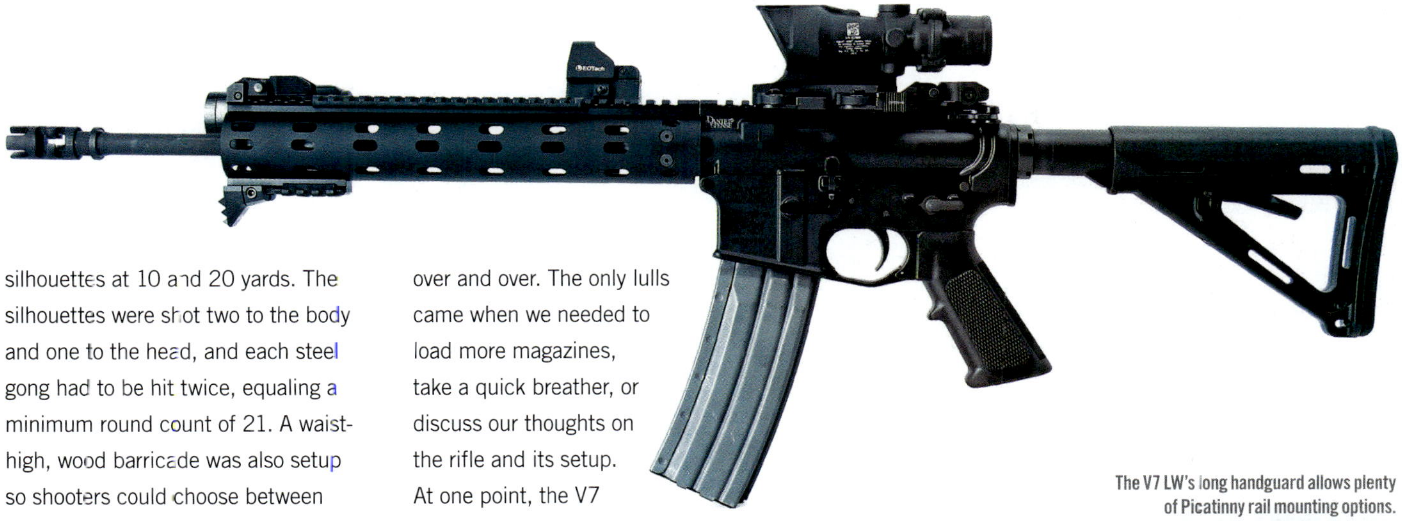

silhouettes at 10 and 20 yards. The silhouettes were shot two to the body and one to the head, and each steel gong had to be hit twice, equaling a minimum round count of 21. A waist-high, wood barricade was also setup so shooters could choose between using the Noveske K9 for support in a kneeling position on the 100- and 230-yard targets, or shooting them standing. Because our personal AR rifles that we regularly run-and-gun with are set up more traditionally with one primary optic, we wanted to see how a two-optic setup would perform in an action-shooting environment.

Engaging the close-range targets with EOTech's MRDS offset in Daniel Defense's 1 O'clock mount was not bad, just different. The carbine needed to be rotated 45 degrees counterclockwise (from a right-handed shooter's prospective) to properly acquire the dot. With repetition, our shooters had no problem getting their hits, transitioning from target to target, and controlling the rifle. We were a bit skeptical about how switching from the MRDS to the ACOG would go, but to our surprise, it was effortless. Even with the TA31H-G's short eye relief (just 1.5 inches) we had no problem picking up the reticle after rotating the gun back to a normal shooting position. Once there, the ACOG's 4x magnification made the long-range steel targets easy pickings.

We ran the V7 LW pretty hard during our course-of-fire evaluation. The rifle was simply passed from shooter to shooter and run through the course over and over. The only lulls came when we needed to load more magazines, take a quick breather, or discuss our thoughts on the rifle and its setup. At one point, the V7 LW's handguard became so hot from the heat coming off the barrel we each had to dawn a glove on our support hand to keep going. Despite the heat we were building up in the lightweight barrel, there were never any problems hitting the 18-inch, steel plates at 230 yards. Granted, we weren't aiming for bullseyes, but just trying to get two hits as fast as possible to move on to the next target. During this test involving firing hundreds of rounds in a short period of time, the V7 LW did not malfunction once; it was as if it was taunting us to try to make it fail.

Daniel Defense's M4 Carbine, V7 LW left us with a great impression. At an MSRP of $1,449 (street price at the time of this writing was in the $1,200-$1,300 range), the price for such a high-quality rifle is hard to beat. With its versatility, the carbine finds itself at home in just about any situation. Its hammer-forged, 16-inch barrel, midlength gas system, 12-inch hand guard, and light weight allow it to be a rifle for all occasions: casual plinking, tactical classes, or 3-Gun competitions; simply outfit as you require. For shooters looking for an all-purpose solution that doesn't break the bank, the V7 LW could be just the solution for you. **R**

The V7 LW's long handguard allows plenty of Picatinny rail mounting options.

DANIEL DEFENSE M4 CARBINE, V7 LW

CALIBER: 5.56mm

MAGAZINE CAPACITY: 30-round Magpul PMAG

BARREL LENGTH: 16 inches

OVERALL LENGTH: 36 inches, stock extended

WEIGHT: 7 pounds, 11 ounces

MSRP: $1,449

URL: www.danieldefense.com

FEATURED ACCESSORIES:
Primary Weapons Systems FSC556 Compensator ($100)
Noveske K9 Barricade Support ($49)
Contour ContourGPS Camera ($300)
Contour Picatinny Mount ($30)
Trijicon TA31H-G ACOG Scope ($1433)
LaRue Tactical LT-100 ACOG QD Mount ($125)
EOTech MRDS Optic ($650)
Daniel Defense 1 O'Clock Offset Rail Mount ($41)
SureFire Mag5-60 High-Capacity Magazine ($129)

PRICE AS FEATURED: $4,306

AN AMERICAN ICON

How the AR-15 Became One of the Most Recognizable Symbols of the USA

By Eric Graves

A true national icon, the AR has become the nexus of American innovation, industrialization, popular culture, and freedom. It's carried by service members, law enforcement officers, and armed citizens. Uniquely American, no other weapon is as recognizable a symbol of our military might.

The father of America's rifle is Eugene Stoner. Born November 2, 1922, only a few years after World War I, the inventor of the ArmaLite Rifle grew up in Southern California and worked in the aircraft industry prior to enlisting in the Marine Corps to serve in the South Pacific during World War II. Appropriately enough, he specialized in aircraft ordnance. After the war, Stoner returned to the aircraft industry and became a design engineer with Fairchild Engine & Airplane Corp. It was in Fairchild's ArmaLite division, where he began work on a series of small arms that would result in the AR-10 and AR-15 rifles. Both leveraged innovations in materials and engineering that Stoner gleaned from his wartime service and honed while working on advanced aircraft designs.

Developed in 1955, the AR-10 was chambered in the new 7.62mm NATO cartridge. Although it was tested by the U.S. Army as a possible replacement for the M1 Garand, the Army chose the more traditional M14 design instead. Undeterred, ArmaLite licensed the AR-10's production to Dutch manufacturer Artillerie Inrichtingen. It's from there that the AR-10 found its way into those Brush Fire wars that spread across the globe.

Next, ArmaLite turned its attention to modifying the design to the smaller 223 Remington cartridge. The U.S Air Force was first to adopt the new AR-15, for its Security Police and Air Commandos. The design was eventually licensed to Colt, and a fully militarized variant known as the M16 is what the U.S. Army took to war in Vietnam. The rest is history: millions of examples of the rifle and its many derivatives have entered service with militaries and police agencies all over the world.

Unfortunately, we lost Gene Stoner on April 24, 1997, but Brownells honors his legacy with their meticulously researched reproductions of the AR line, beginning with the AR-10 variants and continuing through semi-auto versions of the M16 family. The AR's space-age looks inspired many designs after it, including a whole line of ArmaLite rifles featuring different operating systems and in different calibers.

The silhouette of the M16 has come to define American military power. What's more, in its AR-15 form, it's the most popular civilian rifle in America. It also looks great on screen. Over and over again, moviegoers have seen the AR play a prominent role in big screen action movies.

For many in the American public, their first encounter with the new M16 rifle came in a movie near and dear to many a RECOIL reader's heart, *The Green Berets*, starring John Wayne. Unfortunately, it figured prominently in one scene where the Duke broke a rifle across a tree, solidifying the "Mattel Toy" urban legend that dogged the M16 in its early years.

With a full marketing campaign by Colt, and veterans of the Vietnam War returning home to become police officers, law enforcement agencies began to adopt the M16. Then in February, 1975 the American viewing public began watching the prime time TV drama, *SWAT*. Inspired by the Los Angeles Police Department's Special Weapons and Tactics Team, the show's characters carried M16s.

Art repeatedly imitated life with a slew of war movies memorializing America's involvement in the Vietnam War. *Apocalypse Now*, *Platoon*, and *Hamburger Hill* seared the image of the M16 into another generation's brains.

Now a classic, 1986's *Heartbreak Ridge*, starring Clint Eastwood, took us inside a Marine Recon platoon and their short, but intense, combat on the island of Grenada. It was our first major military action since the end of the war in Vietnam and we made quick work of Grenadian and Cuban forces occupying the island.

No one will ever forget the famous line, "Say hello to my little friend," from Brian De Palma's *Scarface*, where a coked-out drug dealer named Tony Montana trains an M16/M203 combo on a team of assassins sent to kill him. The rifle took on the role of costar if only in that one scene, making the line an oft-repeated meme in the American lexicon.

Aside from the screen, probably the strangest pop culture crossover occurred not long after the U.S. Army began to field the M16A1. In order to make maintenance updates more digestible for conscripts, the Army enlisted the services of comic book artist Will Eisner to create a monthly illustrated magazine called "PS, The Preventive Maintenance Monthly." It was complete with an Ol' Sarge character named Master Sergeant Halfmast, along with a buxom lass, with the Bondesque moniker Connie Rod, who was sure to keep the troops' attention. Like the content, the name was inspired as a postscript to maintenance manuals. In addition to tips in the monthly magazine, the Army released "The M16A1 Rifle (DA PAM 750-30)" in 1968 as a one-shot issue, concentrating on rifle maintenance and use. Printed in the millions, it's still available as a collectible. Like the cover says, "Hey Sarge, get this copy to the man with the rifle!"

But it was not until the sunset of the so-called "assault weapons ban" in 2004 that the AR-15 became America's rifle. Prior to the ban's enactment in 1994, few companies manufactured AR clones. Once Congress passed the legislation and it was signed into law by President Clinton, there was a brief period before it came into force. During this period, there was a buying frenzy of what were referred to as "pre-ban" guns of all makes, but it was nothing like what we've seen in panics of the ensuing years. Much of this panic buying was fueled by scarcity and also a lack of information in those pre-internet days.

In 1994 the fabled TDP, or Technical Data Package, essentially the blueprints for the M16 was tightly controlled by Colt and hadn't been leaked yet. The few rifles on the market used commercial parts that weren't manufactured to Mil-spec.

During the ban, black rifles, as ARs came to be known, were available in modified forms, but in order to be eligible for commercial sale, they could not be configured as assault weapons. "Assault weapons" were semi-automatic rifles that had detachable magazines and included two or more of the following features:

- Folding or telescoping stock
- Pistol grip
- Bayonet mount
- Flash suppressor, or threaded barrel designed to accommodate one
- Grenade launcher

Firearms manufactured for sale to law enforcement and the military did not have these restrictions. Consequently, development of these firearms was driven almost completely by government requirements.

Although laws with provisions similar to the AWB remain in place in several states, 2004 saw the end of the law on the federal level, and development and production of ARs skyrocketed. Subsequent calls by lawmakers for a renewal of the AWB have fanned panic buying that has contributed to the AR becoming the most popular rifle in the United States. This increase in demand also fueled innovation with the introduction of new materials, parts, and calibers.

Although it was originally designed by Gene Stoner for military applications, the AR and its many variants are used every day for work as well as recreation, not only here in the United States, but all over the world. It also serves as a symbol of both individual and national freedom. R

GREATEST OF

Michael Jordan is the greatest basketball player of all time. For those who grew up in the '90s this statement was gospel. Nowadays, this fact is constantly debated. LeBron James fans spout numbers and throw around percentages to give their argument validity. When looking at the statistics, the case against Jordan is strong, but greatness cannot be measured in the stats column alone.

What's the metric for greatness? Is it about the accuracy of the shooter or is it more than that? Who'd win in a one-on-one game? Sure, LeBron does things MJ didn't do, but the game was different then. Besides, the match-up could never happen, being that these game changers are different ages. It's comparing apples to older apples that had to play rougher defense. Even though his tongue was out half the time, Jordan affected the game beyond measure, and LeBron is the evolution of that game. It would be safe to say, without Michael Jordan there would be no LeBron James.

Debating the relationship between these two athletes was sparked by RECOIL's latest assignment. The AR-10 is a revolutionary rifle design that influenced the generations of guns that came after it. The AR platform has changed significantly since the 1950s, but the AR-10 still plays a big part in its heritage. Without the AR-10 there would be no AR-15.

HISTORY OF THE GAME

George Sullivan founded The ArmaLite Corporation in the 1950s. The small arms engineering company focused on producing concepts rather than manufacturing. One of the best pieces of firearm industry lore is the story of how Sullivan met Eugene Stoner. While testing a prototype at a local shooting range, Sullivan sparked a conversation with the small arms inventor who he would later hire as ArmaLite's chief designer. The rest is black rifle history.

The ArmaLite AR-10 was a 7.62x51mm battle rifle developed by Stoner. It used a direct impingement gas operating system, composite materials, and a straight-line barrel and stock design. By today's standards, nothing special, but in the 1950s it was a space gun. The company eventually sold a manufacturing license to the Dutch arms manufacturer, Artillerie Inrichtingen (AI). There were many product improvements during the time AI produced AR-10s, resulting in at least two different versions of the rifle. As with many things ahead of their time, the AR-10 was not a big seller in its infancy

ALL TIME

How Does a Modern .308 Compare to a Retro-Inspired Version?

By Nick Saiti
Photos By Kenda Lenseigne

(allegedly less than 10,000 total).

Fast-forward 60 years and the AR-10 is more popular than ever. Different calibers, new technology, and other innovations have created a plethora of options for the platform. The originals have been deemed unicorn guns, and if you're lucky enough to even find a select fire version of the AR-10, it would set you back somewhere around eleventy-million dollars. Totally worth it.

Don't worry kids, Brownells is here to save the day. The Retro Rifle™ line uses brand new, modern materials to recreate Stoner's classic designs. A line of rifles that you don't need to be an archaeologist to find and you wont feel guilty during mag dumps.

JORDAN

The BRN-10 by Brownells is a resto-mod version of Eugene Stoner's original lightweight AR-10 battle rifle. It's meant to revive the nostalgia of the '50s game changer using modern manufacturing methods. It features machined 7075 T6 aluminum receivers, so the upper and lower fit neatly together with almost no play. The lack of a forward assist and shell deflector give it a minimalistic look. The early trigger-style charging handle is located under the carry handle. The lines of the original are definitely there.

The outside is all retro, but under the hood it's all modern. The chrome-plated bolt is magnetic particle inspected to make sure there are no microscopic imperfections. I haven't researched it thoroughly, but I assume this was not a thing in 1956. The 20-inch barrel has a nitride finish with a topcoat that replicates the classic phosphate look. The rifle comes with one DPMS/SR-25 style metal magazine; unfortunately Magpul P-mags won't fit. Many parts are interchangeable with modern AR-15 and DPMS/SR-25 pieces, but upgrading would defeat the purpose of this rifle.

Brownells makes two different versions of the rifle. The BRN-10A mimics the earlier "Sudanese" rifles with its open three-pronged flash hider and brown furniture that looks like the original fiberglass. The 10A weighs in at eight pounds and nine ounces, and is the version we're covering in this article. You can tell the BRN-10B takes its inspiration from the later "Cuban" rifles by the black furniture, closed-prong flash hider, and lightweight barrel.

Machined receivers, 7075 T6 aluminum, nitride finishes, and magnetic particle inspected bolts would have sounded futuristic decades ago, but they're all now standard features. It's like having fuel

injection, electric windows, air conditioning, and a killer sound system in a 1964 Ford Mustang.

LEBRON

Evolution took hold of the AR-10 and has resulted in the Patrict Ordnance Factory (POF) Revolution. The winner of the Golden Bullseye Award and the Industry Choice Award for rifle of the year is the latest iteration of the AR-10 battle rifle. The Revolution is a .308 rifle that handles like a 5.56. Weighing in at 7.3 pounds, the Revolution is not just a cut up version of a .308, but an AR-15 that fires .308.

The Revolution features an adjustable gas piston system that doesn't allow gases to foul the bolt carrier group. The fully ambidextrous billet lower receiver houses a single stage match trigger. The 16 1/2-inch barrel is tipped with a three-port compensator. The barrel features POF's E-squared chamber. Four channels are cut into the walls of the chamber to assist in extraction. POF's oversized heat sink barrel nut offers seventeen times more heat dissipation than a standard barrel nut.

The list goes on with innovations over the standard. There is a reason the Revolution keeps winning awards. It is the culmination of innovations in engineering and technology that push the boundaries of what is possible with the .308 AR platform.

IT'S GAME TIME

Let me take a minute to preface the range report part of this article by saying iron sights coupled with full-house .308 rounds are not optimal for shooting tight groups. End public announcement.

The BRN-10 shot softer than expected. The rifle length gas system helps tame the big .308 round. The gun ranks high on the fun scale. It's not a practical choice, but neither are many good things in life. The first round fired through the BRN-10 was followed by an unfamiliar ringing sound. It took a few rounds till I figured out that the open pronged flash hider rang like a tuning fork. This especially happened when shooting from the hip 1980s action movie style. I just couldn't help myself. The A2 style sights and the 20-round metal mag brought back many childhood memories. It's just too much nostalgia for one man to take.

The gun is well-balanced and very easy to maneuver. The handguard is the perfect size to wrap a hand around. The trigger is pretty good out of the box, breaking at around 5 pounds. In all, it's an able weapon that can do everything it was intended to do. That doesn't sound flashy or fun, but when coupled with the retro look and feel it becomes more than the sum of its parts. Slam dunk!

The POF Revolution didn't shoot that much differently from any other AR-15 I have in the safe except for the fact that the bullets are three times the size. How'd they do that? The gun is lighter and flatter than the BRN-10, but that's to be expected considering the 60-year age difference in the designs.

For accuracy testing both guns were shot from a rested and bagged position at 100 yards. Three different kinds of loads were used: Hornady SST Custom Lite 125 grain, PMC Bronze 147 grain and Federal Gold Medal 168 grain. Wolf Ammunition 145 grain was used for general range shenanigans. The POF shot its tightest group at just over 0.6 inch, while the BRN-10 grouped at just under 2 inches. The lopsided results can be attributed to the lack of magnification on the BRN-10. One can only assume that duct tape is not as effective as scope rings for mounting optics.

The time it takes for follow-up shots is increased without controlling muzzle rise. The compensator on the POF does wonders for taming the muzzle of the portly round, while the flash hider of the BRN-10 simply doesn't do much in this department. Muzzle device and compensator designs have evolved almost as much as the rifle itself.

In tests of speed and accuracy it's tough to beat the POF. This modern battle rifle is built to rapidly and accurately put shots on target. The BRN-10 was built with the same goal in mind, but is understandably a bit less efficient at doing so.

THE BUS RIDE HOME

The goal here was not to show that the newer version of the rifle is better. That would be like saying LeBron is better than Jordan, and we already covered that. The idea is to look at the past to see how far we have come today. It felt surreal holding the gun that started it all and its revolutionary great grandson at the same time. There is something to say for bridging that wide generational gap in just a few hours on the range. The POF is all business and the BRN-10 is a little bit of rock and roll. While the BRN-10 doesn't perform as well as the modern rifle, it's one great weapon with a ton of character. How would you measure greatness? Please don't stick your tongue out while shooting either of these rifles. R

MAKE / MODEL / PRODUCT ID# / PRICE

Brownells
BRN-10A™ Retro Rifle, .308 Win/7.62 NATO 20" Bbl
078-000-419
$1,700

IF RIFLES COULD TALK

Old School Flashback From the Jungles of 'Nam

By Nick Saiti | Photos By Kenda Lenseigne

When you're a solider and don't want to die in combat, you need a rifle you trust, especially when the sh*t's coming at you fast and horizontally. For John Plaster, that rifle was the CAR-15.

A Green Beret staff sergeant, Plaster got to Vietnam in 1968. He was assigned to the Studies and Operation Group (SOG), which despite the academic-sounding name was in the business of kicking ass. And business was damned good, in Laos, that is, where Plaster and his Green Beret comrades ran daily missions to play havoc with North Vietnamese men and materials moving down the Ho Chi Minh Trail.

Plaster's weapon of choice was the 5.56mm CAR-15, officially designated the XM177E2. It was a shorter version of the M16A1 and was developed for Spec Ops teams in need of a lighter, more maneuverable weapon for covert missions. Through numerous firefights and many close calls, Plaster's CAR-15 got him through the worst of it, alive and intact.

"A major benefit of the CAR-15 was its short length," Plaster told me. "Making our way through thick jungle and constantly manipulating our way through vines and bushes, the CAR-15 was more likely *not* to become entangled compared to an M16. Our engagements were close-range and intense — so this size difference truly mattered."

He continued, "I think most SOG veterans are sentimental when it comes to their CAR-15. Even today, a half-century later, I still remember my CAR's serial number: 905442."

AN HOMAGE TO THE PAST

While it isn't a carbon copy of Plaster's favorite lifesaver, the new Brownells XBRN177E2™ Carbine is a pretty faithful reproduction of Plaster's CAR-15. Chambered in 5.56 NATO, the rifle features a 12.7-inch barrel and the old-school M16 carry handle complete with A1 iron sights. It also has a polymer CAR two-position buttstock, and a retro six-hole, rounded handguard — both replicating those found on the original.

Of course, there are differences, the big one being the original CAR-15 could switch to full-auto rock 'n' roll. It also featured a moderator/suppressor at the end of the shorter barrel, while the Brownells version sports a pinned and welded flash hider with no suppression capabilities. The flash hider brings the total barrel length to 16 inches, avoiding the whole NFA Tax Stamp hassle. The ejection port cover on the XBRN177E2™ is an A2-style cover, not the original's A1.

My plan on Day One with the XBRN177E2™ was about getting those open sights aligned with the bullseye. Actually, it became a learning curve about twist rates and bullet choices for the original CAR-15.

PERFORMANCE

I started off my shooting of the CAR-15 with Black Hills Ammunition's 223 Rem loaded with a 52-grain match hollow point bullet. My first shots at 50 yards were on paper, and on a horizontal line with the bullseye — but 8 inches to the left. Easily fixed.

For sights, the XBRN177E2™ employs an elevated A1 front post and an A1 flip-up aperture rear sight embedded in the back of the carry handle. The rear sight is adjusted by depressing the tiny pins on the right side of the carry handle with something pointy (the tip of a 223 round works fine), while a screwdriver in the left-side slot turns the sight in the required direction.

A handful of adjustments and several shots later, the XBRN177E2™ was pegging nice groups on the bullseye at 50 yards.

I loaded the 20-round magazine (included) with Reaper Outdoors 5.56x45mm firing a 77-grain Sierra OTM bullet and let off 10 shots at another 50-yard target. And was shocked to see holes all over the target in an 8- to 10-inch circle, with none even *near* the bullseye. What the hell?

At the target, I saw the problem. The 77-grain bullets were tumbling, smacking the paper damn near sideways. So, I tried Black Hills 223 Rem 62-grain Match King Hollow Points. While not nearly as unbalanced as the heavier Reaper bullets, these 62 grainers were still drilling through the paper at an angle — and generally avoiding the bullseye, too.

Paul Levy, director of product management for Brownells, told me the rifle was made as closely as possible to the original. This included using the original rifling twist, 1:12. That numeric designation refers to how fast the rifling inside a barrel does a full twist; the 1:12 twist does a complete revolution every 12 inches.

However, to stabilize longer, heavier bullets, a faster twist rate is required. Most of today's AR-15s are rifled with a twist rate between 1:7 and 1:9. The U.S. military's M4s are 1:7. But the XBRN177E2's 1:12 is just too old school to put an accurate spin on any of the bullets I shot weighing over 55 grains. I asked Plaster if he'd had any problem with bullets tumbling or other accuracy mishaps with his CAR-15.

"We zeroed our CAR-15s at 100 yards, at which distance they shot groups of roughly 1.5 to 2 inches, using 55-grain standard military ball ammunition," he said. "That's not pinpoint accuracy, but the great majority of our firefights took place at well under 100 yards, due to the heavy jungle we fought in."

After some trial and error, I settled on three rounds to test the XBRN177E2™ for accuracy: American Eagle MSR 5.56 NATO and a 55-grain FMJ bullet; Black Hills 223 Rem with a 52-grain match bullet; and Team Never Quit's 223 Rem training round featuring a 45-grain frangible bullet.

I also decided the 100-yard accuracy testing with the iron sights was only going to prove that my middle-aged, glasses-wearing eyes may need a new prescription. So, I attached a scope to the XBRN177E2™, a Nightforce NX8 1-8x24mm, mounting it atop the rifle's carry handle with a Brownells AR-15/M4 Carry Handle Mount.

It was an awkward set up, as carry handle and scope mount put the optic way too high for a cheek weld on the stock. I got it steady, though, with a sandbag rest, and I know my groups were tighter than they would have been otherwise.

For accuracy, the Black Hills 223 Rem averaged the best five-shot groups at 1.59 inches. Overall, the groups here went from 1.30 inches to 1.90 inches. The American Eagle 5.56 had the single best group at 0.934 inch, but the average was much larger at 2.28 inches.

Team Never Quit rounds averaged 3.34-inch groups of five shots. That poor showing surprised me. When I was trying out different varieties of 223 Rem and 5.56 to find ones that didn't tumble, the Team Never Quit rounds pegged a 1-inch group at 50 yards. Why did the groups expand so drastically at 100 yards? Rifle doesn't like the ammo or vice versa is the easy answer. I've used the TNQ ammo with other AR-15s and recorded MOA results.

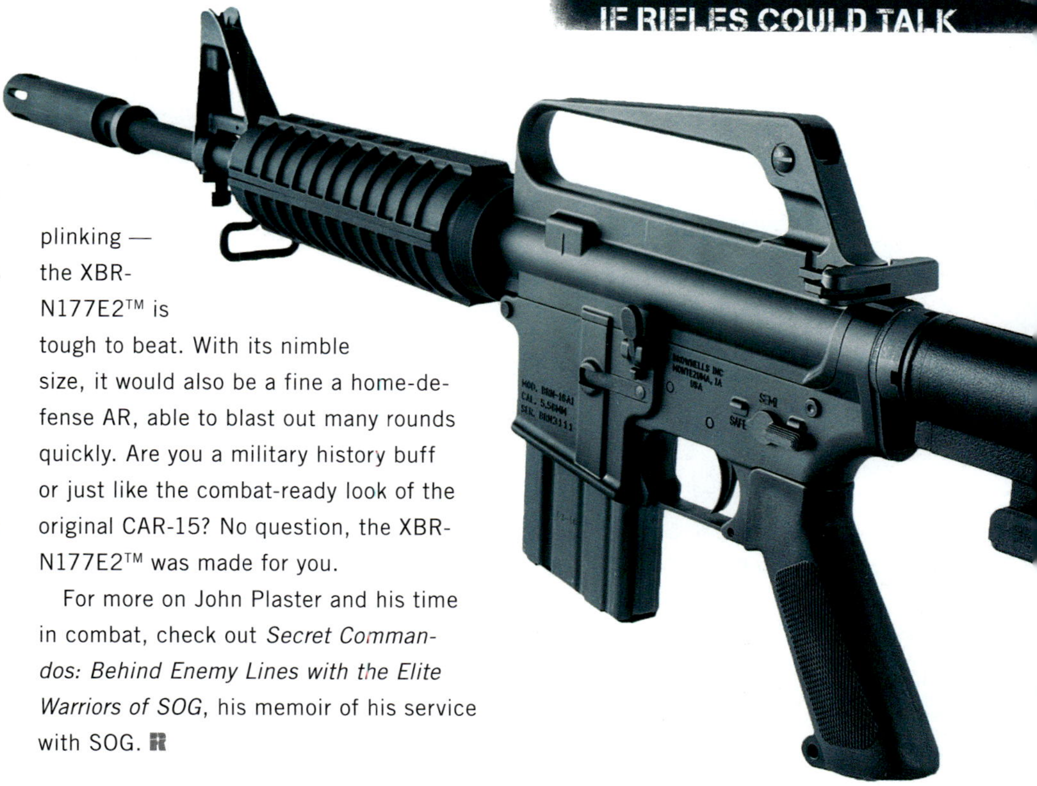

PERSPECTIVE

I'll take some of the credit for the largest groups, but I think the reality of the XBRN177E2™ and accuracy is this: if you find an ammo punching groups at 2 inches or under, that's about the best this rifle will do, given the twist rate.

The XBRN177E2™ has an average Mil-spec trigger: some grit and an uneven pull, but it did start to smooth out after 200 rounds. My Lyman Digital Trigger Pull Gauge measured the trigger's pull weight at an average of 4.5 pounds. When shooting fast and furious, the XBRN177E2™ heats up. But the handguard does a good job of protecting your hand, while the rifle itself cools off fast.

For a sharp-looking rifle that's a lot of fun to shoot — at the range and just plinking — the XBRN177E2™ is tough to beat. With its nimble size, it would also be a fine a home-defense AR, able to blast out many rounds quickly. Are you a military history buff or just like the combat-ready look of the original CAR-15? No question, the XBRN177E2™ was made for you.

For more on John Plaster and his time in combat, check out *Secret Commandos: Behind Enemy Lines with the Elite Warriors of SOG*, his memoir of his service with SOG.

XBRN177E2™ Carbine By the Numbers

Action Type: Direct Impingement	Front Sight: A1	Rear Sight: A1
Barrel Length: 12.7 inches	Overall Length: 34 inches	Stock Material: Polymer
Capacity: 20+1	Magazine Included: one 20-round	Weight: 5.3 pounds
Cartridge: 5.56mm NATO	Magazine Type: Removable	Country of Manufacture: USA
Finish: Black	Muzzle: Flash Hider	MSRP: $1,300

Ballistic Data	American Eagle MSR 5.56x45mm 55-grain FMJ	Black Hills Ammunition 223 Rem 52-grain Match Hollow Point	Team Never Quit 223 Rem 45-grain Frangible
Average velocity	2,849.7 fps	2,685 fps	2,813.6 fps
Standard deviation	37.9	22.7	31.4
Average group (100 yards)	2.28 inches	1.59 inches	3.34 inches
Best group (100 yards)	0.934 inch	1.30 inches	2.08 inches

** Velocity figures are averages of 10 shots recorded by a PACT Professional XP chronograph. Groups are the result of five 5-shot groups from a sandbag rest at 100 yards.

MAKE / MODEL / PRODUCT ID# / PRICE
Brownells **XBRN177E2 Retro Rifle™** 078-000-403 $1,300
Nightforce **NX8 1-8x24mm Scope w/FFP FC-MOA Reticle** 524-000-393 $1,715
Brownells **AR-15/M4 Carry Handle Scope Mount** 080-000-660 $30
American Eagle **5.56x45mm NATO 55gr XM193 Case** 105-000-601 $171 (450 rounds)
Black Hills **Remanufactured 223 Remington 52gr Match Hollow Point** 105-001-424 $31 (50 rounds)

TREAD HEA

Sig Sauer Steps Up With Its Entry-Level AR

By Rob Curtis

Another day, another threat to the Second Amendment. Welcome to the post-Sandy Hook political cycle where left-leaning candidates for office play a perpetual game of gun control oneupmanship. America's favorite rifle, the AR-15, is back in the spotlight as a million-rounds-per-minute ghost gun that gives ER doctors PTSD. While we don't give points in our reviews for marketing prowess, we have to point out that Sig Sauer's naming its lowest-priced AR "Tread." A reference to the Gadsden flag is a fine way to embrace the idea that the AR is both a symbol of freedom and a tool of self-defense.

The M400 Tread is Sig's entry-level AR, and while it's not the cheapest AR-pattern rifle on the market, its design, materials, features, and quality of its components set a value versus performance baseline for what an AR should be as we approach the platform's 60th birthday.

GARANIMALS AND TAINTS

You can build your own, but unless you're an accomplished armorer, you're bound to spend as much as, if not more than, the cost of the Tread by the time you figure out what AR components play nicely together. And, after weeks of workbench time, your build probably won't have all the features of the Tread. And, most importantly for a novice AR owner, it won't be warranted to run.

Don't get us wrong — there's absolutely nothing wrong with building an AR from components. There's no better way to learn about the ins and outs of the platform. But, at risk of being beaten to death with an armorer's wrench in the aisles of SHOT Show, we recommend a build project as a second rifle. Getting an out-of-the-box shooter as a first rifle gives you a chance to decide what you like and what you don't. With that knowledge in hand, you can either start swapping parts on a known, working rifle ... or start a component build. Plus, using a running gun as a baseline means you've got a lot less to troubleshoot when that new trigger doubles or fails to ignite a primer.

Or you could start out with something like the Tread and take advantage of the company's modular approach to upgrades and accessories. Instead of leaving its freshly minted AR owners alone in the jungle of unfamiliar specifications and abbreviations, the company released a line of made-for-Tread, factory upgrades that are guaranteed to work.

Choosing compatible parts for the Tread is as simple as finding the Gadsden yellow logo on the box of a Sig accessory. Sig's EVP of Commercial Sales and Chief Marketing Officer Tom Taylor said, "It's like Garanimals for guns."

Instead of getting the runaround of a trigger manufacturer blaming a

Stock parts (lower left) are easily upgraded with the parts shown: Tread 15-inch handguard, Tango6T optic, Tread Romeo5 red-dot sight, Tread ambi charging handle, Tread 3-chamber brake, and Tread flip-up sights.

failure to fire on an out of spec lower receiver, Sig built the whole damned gun, knows all its parts inside and out, and will help you get it running or fix it themselves if you can't.

Before we give the wrong impression, though, we've run a couple of Tread rifles through a couple thousand rounds over the course of a few months without a single stoppage or mechanical failure.

Bringing this back around to value, we recognize the Tread costs more than rifles at the distal end of the performance and reliability scale, but its price is on par with its competitors in the AR sweet spot, the so-called "Taint" price point.

We're calling it the Taint space because $700 to $800 rifles aren't dirt-cheap, nor are they comparable to $1,300 to $1,800 feature-laden AR rifles. Or maybe we're indulging our inner 13-year-old and embracing the unintended combination of Sig's Tread branding with that of the competing Springfield Armory Saint line of rifles that occupies a similar market space. Either way, Sig provides a lot of value in the Tread for its $800 street price.

Running the gun in its out-of-the-box configuration for several 200-round training sessions over a few weeks set a solid baseline for comparison to the more tricked-out Tread we'd end up with after accessorizing it with Sig's Tread-labeled upgrade parts.

UP FRONT

The first thing we noticed was the slim handguard. The stock M-LOK handguard gets points for ergonomics. Average-sized hands will feel at home with a C-clamp, thumb-over grip or a conventional support-hand cradle. All three Tread handguards have the same profile and omit the top rail, which adds comfort. However, the missing top rail is a drag for guys who want to run a top-mounted aiming laser, such as an ATPIAL or a clip-on night vision setup. Luckily, Sig offers sight-height, M-LOK rail sections to mount irons, lasers, and anything else that'll fit on a 2- to 3-inch rail section.

We ran the rifle pretty hard in a positional drill that required dumping 15 rounds in prone, kneeling, then standing positions. Five reps and 80 rounds later, the heat from the chamber passed through the barrel, into the unvented portion of the handguard where it built up and crept forward. It was hot enough to call for shooting gloves, though not hot enough to stop shooting while barehanded. The 15-inch accessory handguard ran noticeably cooler with its greater number of M-LOK slots, allowing heat to escape.

Swapping the handguard is a two-bolt operation and takes two minutes if you're drunk or blind, or both; otherwise, it takes maybe a minute. The setup is similar to Heckler & Koch's 416 channeled nut and clamping handguard system. It affords a long mating surface for the nut and handguard, reducing the diving board effect seen with shorter barrel nuts. The Tread is equipped with Sig's micro gas block that accommodates the slimmest of handguards — though your choices are limited to Sig's at the moment, since the Tread uses a proprietary barrel nut.

It took a while, but we did find an issue with the Tread that was com-

Sig's Tread-branded flip-up irons are all-metal, lock in the up position, and include the sight-height M-LOK rail section to mount the front sight at the proper height.

mon across all three handguards (stock, 13-, and 15-inch models). They didn't flex, shift, or fail — they drew blood. At some point while working rifle-to-pistol transitions, the handguard's rear edge (or the ears that wrap around the upper receiver) caught and left a jagged scratch on our forearm. These areas aren't dehorned on any of the Tread handguards, and they should be.

CONTROLS

The stock trigger surfaces are all DLC-coated for improved wear and feel, but there's no hiding the heavy, 7-pound, 11-ounce average pull. The best way to describe the break is like popping industrial grade bubble wrap. We're talking about the kind with the ¾-inch bubbles. Shooting groups with it demanded a lot of concentration, particularly because we found a tiny hitch in the take-up when shooting very slowly. We were staging the single-stage trigger about a tenth-of-a-millimeter into the pull to find the actual wall before the break. Shooting at speed, it's unnoticeable. It did make us appreciate Sig's optional, single-stage Tread upgrade trigger.

Swapping triggers was as easy as swapping any other non-cartridge trigger — perhaps a little easier since the trigger included a slave pin to hold the disconnector in the trigger body while inserting the trigger pin. It also came with a coned hammer installation pin, making it easier to center the hammer pin before driving it all the way through the receiver. The flat faced, single stage trigger registered a very crisp 3-pound, 9-ounce break. The take-up is long, but absolutely clean with a crisp break rivaling any aftermarket trigger out there. Anyone buying a Tread would do well to consider this as the rifle's first upgrade.

The ambidextrous, 90-degree safety has a solid, if slightly heavy, feel, and is shorter on the right side to give the trigger finger room to work. It's easily knuckled backward to safe. The ambi mag release has good leverage when run from the left side, and the angled bolt release prevents its inadvertent activation during vigorous paddle-slapping reload maneuvers.

INTERNALS

The carrier key, yawn. Yes, it's properly staked, as is the castle nut out back. We expect that from Sig.

But what we weren't expecting was the extractor support in the barrel extension. We consider this a high-end touch and one that Sig uses in its 516 series rifles. When the bolt of an AR is in battery, the extractor is an unsupported area of the bolt head. In an overpressure event, the hinged extractor acts like

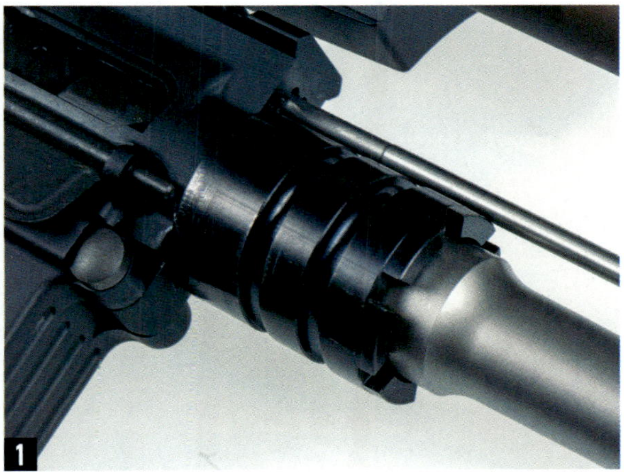

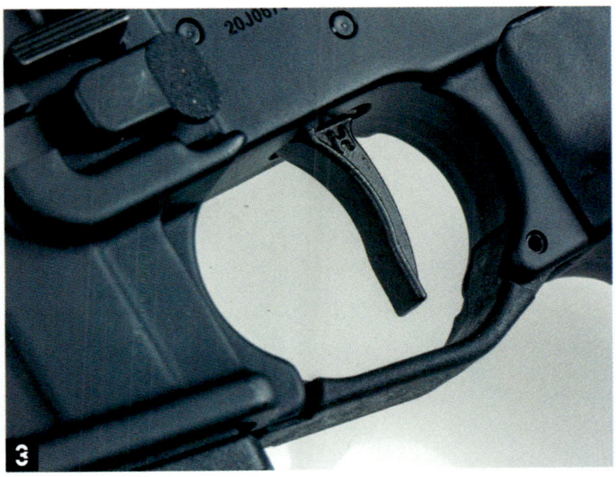

1. Sig Tread's proprietary barrel nut.

2. The edges of all three Tread handguards are sharp enough to warrant deburring.

3. Tread's single-stage upgrade trigger features a flat-ish trigger shoe and a 3-pound, 9-ounce break.

4. Tread's out-of-the-box ambi controls include a left-side mag release and an extended, angled bolt release.

5. This polymer wedge tightens up the upper to lower fit.

The tread's 1:8 twist barrel includes a tapered shoulder that reduces concentricity issues when used with tapered muzzle devices.

a relief port and lets gases blow it open. Sig places a pin in the barrel extension that essentially braces the outside edge of the extractor, preventing it from opening when the bolt is in battery. It's a great safety feature providing insurance against a receiver blow-out if you suffer an obstructed bore malfunction.

In the lower, Sig includes an Accu-Wedge (or something that looks and works identically) to take up any slop between the upper and lower receiver. We also noticed the Tread lower is the same forged lower used in all of Sig's AR pattern rifles. It sports thick walls and QD sling sockets on either side at the base of the receiver extension tang. In use, the sockets place the sling's ring ends just far enough from the hand to work well in most positions, but running the rifle when slung this way doesn't feel natural if you like to shoot with tension on the sling, as we do.

MANUAL OPERATION

Odd one, but the manual is pretty helpful for novice AR owners. It calls out all the rifle's components and explains the cycle of operation, showing how to take the gun all the way down to the extractor spring with very clear diagrams. Anyone with a box of take-off A2 grips and birdcage flash hiders would scoff at this, but we all started somewhere. If the amount of effort Sig put into making the M400 owner's manual is any reflection of how much effort went into the gun, then we expect few cut corners.

DOWN TO BUSINESS

The 16-inch barrel is made of 416R stainless steel and runs a mid-length gas system with 1:8 rifle twist. The recoil impulse is manageable, and the gun is easy to control, even easier with Sig's three port brake upgrade.

We've put about 2,000 rounds through a couple of Tread rifles in a variety of configurations from bone stock to gussied-up with all the Sig-branded upgrades available, including the Sig Optics Romeo5 red-dot sight and the Tango6T vari-

The stock Tread configuration upgraded with Sig's Tread Flip-Up Irons Sights and Romeo5 red-dot optic.

above it. Predictably, the faster barrel twist rate favored Federal's long 77-grain Sierra Match King loads, but it did surprisingly well grouping Black Hills light 52-grain loads. Oddly, though, the Tread basically vomited Hornady's 73-grain ELD-Match rounds, which are generally good performers.

	Average FPS	Extreme Spread	Standard Deviation	Group in inches @ 100y
Sig Elite 55gr Ball	2,279	57	257	1.21
Black Hills 52gr Match HP	2,908	36.2	122	0.91
Federal 77gr SMK BTHP	2,445	10.6	33	0.74
Hornady 73gr ELD-Match	2,568	24.2	79	3.8

able power optic. We ran 100- to 200-yard drills with both, shot our accuracy test groups, and moved back to 400, then 500 yards with the Tango6 before we ran out of range. The rifle performed reliably and admirably. We suffered zero malfunctions with this platform.

Accuracy is on par with any AR in the Tread's price range, and many

LOOSE ROUNDS

With a million choices, it's tough to get excited about another AR. But Sig's M400 Tread bridges the gap between performance and value, while adding important features for AR snobs like the extension support pin. Whether you're a card-carrying member of the *No Step on Snek* crowd or just looking for one of the best values in AR-pattern rifles, the Tread should be at the top of your list. **R**

FEATURED ACCESSORIES:
Sig Optics Tango6T (61231) - $1,680
Tread 15-Inch M-LOK Handguard - $149
Tread Single Stage Flat Blade Trigger) - $119
Tread 3-Chamber Compensator - $49
Tread Ambidextrous Charging Handle - $49
Tread M-LOK Forward Grip Kit - $49

OTHER TREAD ACCESSORIES:
Tread Romeo5 RDS - $180
Tread 13-Inch M-LOK Handguard - $149
Tread Adjustable Flip-Up Sights - $149
Tread Sight Rail - $29

PRICE AS FEATURED: $3,045

SIG SAUER M400 TREAD

CALIBER: 5.56mm
MAGAZINE CAPACITY: 30 rounds
BARREL LENGTH: 16 inches
OVERALL LENGTH: 32.5 inches
WEIGHT: 6.9 pounds (8.45 pounds, as shown)
MSRP: $950
URL: sigsauer.com

2

Weird Calibers

BELT-FED. 9MM. UPPER.

Say what? Freedom Ordnance defies all logic with the FM-9

Story by Iain Harrison
Photos by Kenda Lenseigne

There are certain words that shouldn't ever go together in a sentence. Words like habanero and Astroglide, or Clinton and bathing suit. Until Freedom Ordnance released their FM-9, you might have thought that belt-fed 9mm was one of those combinations, but in this case, the juxtaposition actually works better than expected.

Now, there are those who might believe this to be a superfluous toy, unworthy of consideration. But hold on there, Timmy, 'cause not every gun has to be over-the-beach capable — no one is pretending that this is going to be the next infantry light support weapon. Some guns exist purely for range fun and this one is more than ready to fulfill that role and leaves anyone that shoots it grinning like a Cheshire cat.

When we first heard about it, we were prepared to be underwhelmed. Firearm novelty items tend to suffer from a lack of development — after all, why spend R&D dollars when your customers have low expectations in the first place? Instead, we were pleasantly surprised by both the quality of machining and thought that went into its design. Our first impressions were of a gun that was through a rigorous process of refinement — definitely not something that was haphazardly slapped together in a garage machine shop. Individual parts were nicely machined and finished, rather than looking like they were gnawed out of scrap by an angry beaver. Clearances were tight, and components moved slickly past each other with a minimum of friction. Having had some prior experience running CNC machines, we're always on the lookout for evidence of sloppy tool paths or poor calibration. Neither were evident and about the only worthwhile improvement we could find would be to radius some of the sharp edges on the top cover's perimeter.

While a pistol-caliber belt-fed may seem counterintuitive, there's actually a historical precedent for it. The very first working example of Hiram Maxim's machine gun was a one-third scale model in 7.62x25, which also by definition made it the world's first SMG. While the FM-9 is not as *fin de ciecle* drop-dead gorgeous as Maxim's prototype, it makes up for it by being chambered in a more practical cartridge and has the benefit of utilizing disintegrating links, rather than a cloth belt. Because of its diminutive dimensions, the guys at Freedom Ordnance had to design their own links for the 9mm Parabellum round, adding significantly to the effort needed to bring the upper into existence. Performing the same role as a magazine, i.e., getting a live round into position at the right time to be chambered, links face a short, violent existence. Instead of more or less constant spring pressure forcing rounds against a mag's feed lips, links have to endure being yanked sideways at some pretty severe g-forces, then stopping dead in the space of less than a millimeter. Strain exerted by the gun's feeding mechanism varies according to the length of belt that must be lifted — fire an M240 from a vehicle mount with a loose belt trailing on the floor, and if you listen carefully, you can hear the rate of fire change as the belt shortens.

With a gas-operated gun, the user can typically adjust the amount of gas entering the system to compensate for different ammunition types. As the FM-9 features blowback operation, this option isn't available — instead the shooter can either

Magwell insert (right) provides a fixed ejector and dovetail for attaching 150-round nutsack (opposite page bottom).

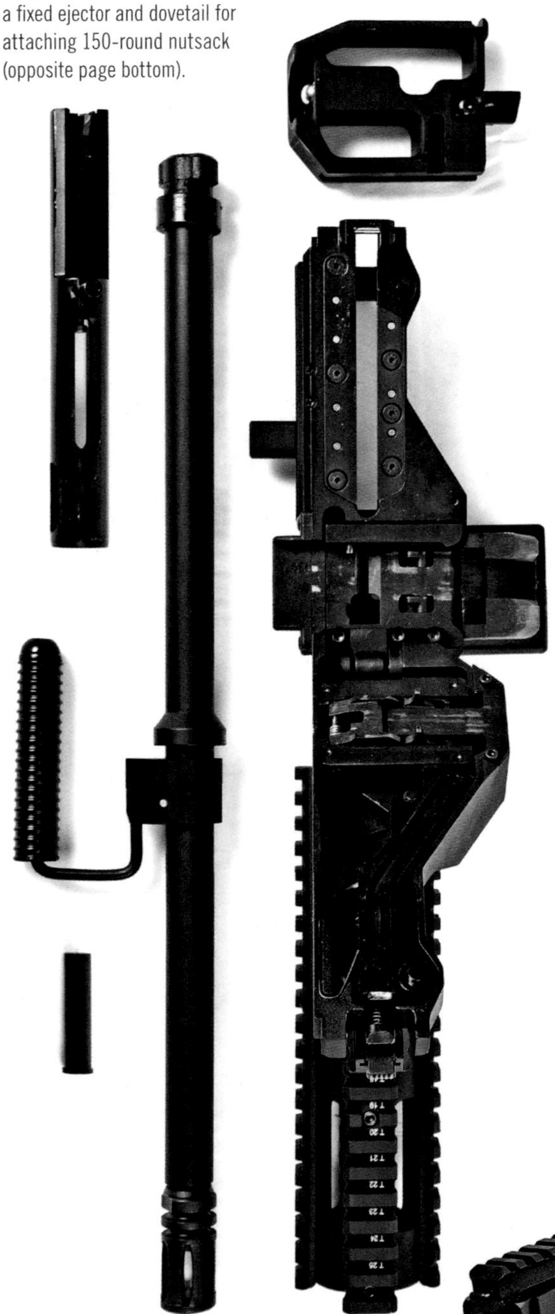

change spring weights or alter the mass of the moving parts. To facilitate this, the bolt features a cavity with a removable weight; if that's not enough to get things working reliably, there's always the option of swapping out buffers. As the bolt cycles, a roller cam pin at the 12 o'clock position engages a track in the top cover, just like in a grown-up machine gun. This track then transfers motion in the gun's X-axis through 90 degrees and uses it to move a pair of feed pawls, advancing the belt and feeding the next round. It's a system that's been successfully used on just about every machine gun since the MG42, only in this case, it looks like the working parts have emerged from a too-hot tumble dryer. Forward of the cam pin lies a feed horn, which rams the fresh cartridge out of its link and stuffs it into the chamber, while the empty link shuffles out the right side of the gun. Also like a grown-up MG, the little nine features a quick-change barrel system — simply depress the barrel retaining latch, and use the carrying handle to rotate the hot barrel to the unlocked position and pull it out of the receiver.

A note on that bolt. Nine-millimeter ARs usually have solid bolt-carrier groups that weigh about the same as a school bus, slam around like a fat guy on a pogo stick, produce way more recoil than a tiny round ever should, and beat on hammer and trigger pins like they didn't listen the first time. The FM-9's bolt weighs a scant 12 ounces, and we were initially concerned that it was going to blow the buffer out the back of the receiver extension due to excessive velocity. We needn't have worried, however, as much of the initial push is taken up by the belt-feed system, effectively producing a delayed blowback action and reducing recoil to almost .22LR levels.

FREEDOM ORDNANCE FM-9

CALIBER: 9mm

MAGAZINE CAPACITY: 150-round bagged belt

BARREL LENGTH: 16+9 inches

OVERALL LENGTH: 25 inches with 16-inch barrel

WEIGHT: 7.2 pounds

MSRP: $1,895

URL: www.freedomordnance.com

FEATURED ACCESSORIES:
Lucid HD7 Gen 3 - $250
Accu-Tac SR-5 Bipod - $235

PRICE AS FEATURED: $2,380

As the FM-9 is not a firearm, it can ship directly to your door. It also has the virtue of being able to snap onto any AR-15 lower, even a full auto version, should you be fortunate enough to have access to one.

ROUNDS DOWNRANGE

If you have any time on an open-bolt MG such as a M249 SAW, M60, or M240B (or GPMG for our British readers), then yanking back the charging handle, raising the top cover, and flipping up the feed tray will probably be second nature. If you try this with the FN-9, you'll quickly run into difficulties, as it's a closed-bolt design. Unless you manually lock the bolt to the rear using the bolt catch, its cam pin winds up in the wrong position to engage the corresponding slot in the top cover. Don't worry though, the gun will tell you that you screwed up, as the cocking handle will only move about an inch before it bottoms out.

Loading is another area where the FM-9 departs from convention. The belt is inserted while the top cover is closed, stuffing links in from the left until a couple of detents snick into place. If you try to lay the belt on the feed tray with the top cover in the raised position, most times the belt sits a little high and produces a first round misfeed. We struggled mightily with this until we actually read the manual and applied some of that book learnin' to our screw-ups. Once you do what the manufacturer tells you to do, things go a lot smoother.

While perusing the manual, we noticed that the guys at Freedom Ordnance helpfully listed a number of ammo types they'd tested, all of which was 115-grain ball. So naturally, we grabbed up just about anything that wasn't on the list and headed to the range. It takes a while to tune the gun to accept various bullet weights, but we managed to get it to run anything up to 147-grain subsonics relatively reliably. When a stoppage did occur, we had to remember not to revert to SOP and, instead, first lift the top cover before pulling the charging handle.

As the upper has the ability to shoot a large volume of ammo quietly and also possesses a quick-change barrel, this got us to thinking (which admittedly, is always a dangerous thing). If the FM-9 were ever chambered in .22 TCM 9R, the user would have the ability to launch a 35-grain bullet at around 2,700 fps and then switch barrels to shoot 147-grain subs — a miniaturized, poor man's .300 Blackout belt-fed if you will. Which would be a whole new level of awesome. R

AMG

ARES DEFENSE MCR

It's Time to Get Your Rambo on

By Peter Palma
Photography by Kenda Lenseigne

For most civilian shooters, the opportunity to shoot fully automatic firearms is as rare as it is fun. Often, when a person shoots a full-auto for the first time, they will deplete the entire magazine in one trigger pull, then get annoyed at their idiot friend who is recording the affair on a cell phone held vertically.

Dumping a mag through an AR or a sub-gun at Knob Creek is no doubt a worthwhile endeavor, but current or former military who were properly trained in the use of machineguns know that this isn't an effective use of the firearm. When the lead is flying both ways, a standard rifle magazine holding 30 rounds can be emptied plenty fast enough on semi-auto — no assistance from an auto sear is required. If you have a machinegun and are using your weapon to suppress a target, it's best to stuff ammo into that pig via a 200-round belt.

Enter the Ares Defense MCR, or Mission Configurable Rifle. The MCR is a belt-fed 5.56mm upper for the AR-15 lower. This ingenious design was meant to be a replacement for the military's M-249 SAW light machinegun. Now, we would drink FN Kool-Aid even if Jim Jones himself poured it for us, but the MCR has so many advantages that we had to have a taste of Ares Defense's concoction.

ARES CAME, ARES SAW, ARES CONQUERED

Upon picking up the MCR, we noticed it was lighter — *much* lighter — than the SAW. At 9.1 pounds with bipod, the MCR came in at just over half of the SAW's mass. This weight saving was dramatically noticeable on a short run to the top of the hill we were shooting from and would undoubtedly be appreciated by anyone carrying it into combat.

Since the MCR uses a standard AR-15 or M-16 lower, the options for stocks, grips, triggers, magazines, and anything else in the vast ocean of AR products become available on a light machinegun. Unlike the SAW, a full-auto MCR lower also has a select-fire option, so the shooter can choose to release a single, well-aimed round at an individual target if a civil war volley of fire wasn't called for. Though the SAW has a welded-up afterthought that will accept the STANAG magazine (and chew the hell out of it), the MCR was designed from the outset to feed from either mags or M27 links.

Like an AR, the MCR upper can be removed from the lower, so the disassembled gun can fit into most backpacks. Pull out the quick-change barrel, and the package becomes downright tiny. The magazine release is in the same ergonomic location and would be a seamless transition for anyone used to the AR platform. The MCR more closely resembles an M-4, effectively camouflaging the machine gunner from being targeted by people who do not like to be shot at by machineguns. Also, the manual of arms is the same since the MCR fires from a closed bolt.

Before anyone goes crying about cook-offs or runaway guns being dangerous, please understand that we have personally seen more negligent discharges on SAWs than on any other weapon. We've seen users ride the bolt home on a round thinking they were closing the bolt, like in training, which resulted in a very loud surprise. We've had a runaway gun when the pin holding on the rear pistol grip and sear fell out while disengaging the safety. It's also not the best of ideas to ride around in a bumpy vehicle where the only thing keeping your firearm from letting a good burst go is a ½-inch sear holding back all the potential energy of the recoil springs.

If this feed tray looks like a SAW, M60 or M240, it's because they all ripped off the MG42. Why reinvent the wheel?

The MCR's forward quad Picatinny rail system allows for the use of better bipods, lights, and optics, and allows them to be swapped out by the shooter as their needs change. However, the handguard is proprietary, so unlike the ubiquitous AR, it cannot be switched out for aftermarket systems. The handguard is also short, so if you want to use a Chris Costa-style grip, you will leave your epidermis cooking on the gas piston. Our sample brandished burnt fingerprints from the previous bearded shooter.

The chrome-lined, hammer-forged, 1/7-twist barrel can easily be removed by locking the bolt to the rear, pushing a button, and pulling it straight out by the carry handle. This aids in cleaning and also allows for rapid barrel changes as they heat up under sustained fire. Each barrel has a removable, short-stroke piston positioned on the left side with adjustments for off, suppressed, regular, and dirty. The newer SAWs have no provision for cleaning the gas port, nor do they have a setting for suppressed fire.

The MCR was well thought out and attention paid to the details. There is no useless forward assist and throughout the weapon there are removable parts cleverly and passively retained by other parts. For example, the piston rod retains the charging handle, and the barrel retains the piston rod. The non-reciprocating charging handle is on the left side of the weapon in a slightly more ergonomic and familiar place than that of the SAW, located on the right.

Both the SAW and the MCR utilize rotating bolts, but the MCR has a wave form bolt in a proprietary carrier, offering greatly improved strength. The belt-feeding mechanism has changed very little from the SAW (almost all NATO machine-gun feed systems were borrowed from the German WWII MG42) and the iron sights are comparable. In order to adapt the mag-fed AR-15 lower to use standard 200-round boxed belts, or the 100-round cloth nutsack (and, yes that is a thing), the MCR employs a machined aluminum magwell adapter, which inserts in the usual fashion. Once a belt is depleted, the shooter can either remove the empty box and insert another or press the magazine release and insert a magazine.

ROCK 'N' ROLL

On paper, the MCR is leaps and bounds ahead of the SAW. But paper and real life are worlds apart, so we wanted to see how it performed on the range in 107-degree conditions. The samples we tested were fitted on registered full-auto AR lowers, but before we reenacted the closing scenes of Rambo II we were curious to see what kind of groups we could squeeze out of this beast. A five-shot group out of the 16-inch barrel at 100 yards using Federal linked ball and iron sights yielded a group size of 2.9 inches, about on par for a military-issue 5.56mm rifle using the same ammo.

Because of its semi-auto mode, the MCR would make much better use of the 4X ACOG the SAW is currently equipped with. In addition to iron sights, this particular MCR was equipped with a small, lightweight Shield CQS mini red dot with a reticle similar to that found on an EOtech. Our first burst on full auto produced what seemed to be a very high rate of fire and a significant amount of muzzle rise. After adjusting the gas setting from "dirty" to "normal," we had a much more manageable rate of fire at 900 rounds per minute (recorded with a shot timer). Anyone familiar with the SAW would be right at home.

The MCR charging handle is non reciprocating and mounted conveniently on the correct side of the weapon (though a lefty might disagree) - if it were made of steel it would be just about perfect. Note burned-off lube coating the gun after a couple thousand rounds of M855. We run them hard.

The MCR proved to be so controllable, it freed up the shooter for other tasks. And yes, this is an actual screen shot complete with typos (but with numbers deleted), not a Photoshop fake. (Note: The shooter confirmed his target and its surrounding before texting and shooting, and did so strictly for review purposes. Not in shot are the two safety officers; don't try this at home, kids.

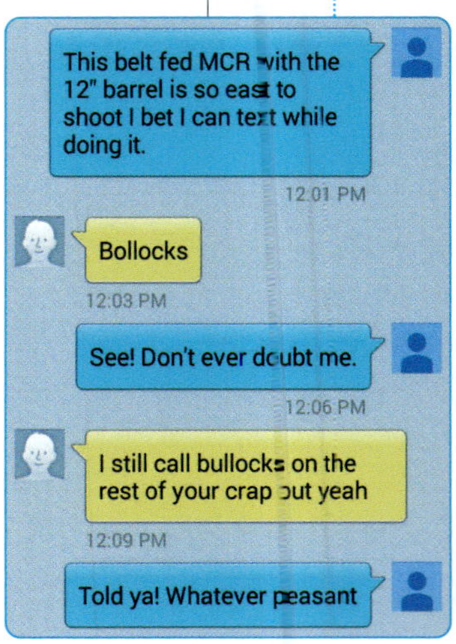

After expending 200 rounds in a series of controlled bursts, it was time to change belts and barrels. Swapping in a second box and a shorter, 12.5-inch barrel, we noticed that the shorter barrel's rate of fire slowed down to 750 rounds per minute and the firearm became so controllable that we fired only a few bursts before our inner 10-year-old took possession of our index finger. Despite using the gun like a fire hose, we still had absolute control over the muzzle and could transition from one target to the next at will, almost as if the gun was not firing. It was then that we had the thought that we could probably send a text while shooting the MCR on full auto. So we did.

We were so amazed at the MCR's controllability that we invited some of our Marine friends over to shoot it. Since Marines break absolutely everything they touch, we figured this would be an excellent reliability test. A couple thousand rounds of Federal M855 green tip were expended, some of which were sent downrange 200 at a time, rather than at the measured cadence of the six to eight round burst that the book calls for.

The only non operator-induced malfunction that occurred was when one of our Marines went into beast mode and sheared off the end of the aluminum charging handle. The charging handle can have immense forces placed on it, especially by Neanderthals, and should really be made from steel. We replaced the charging handle and had no other malfunctions even when firing non-recommended 55-grain ball ammo (Winchester and American Eagle) through both P-mags and C-Products Defense stainless steel magazines.

All of this fun and suppressive firepower comes at a price. So what if you don't have 20 grand laying around for a transferable AR lower? Or the five large for an MCR? Fear not, Ares has solutions.

One of the options open to an average shooter like yours truly, is that the MCR can be purchased with a magazine-fed-only top cover for $2,000 — not much more than for a good-quality gas piston AR. This still gives you the quick-change barrel feature, improved bolt, and the ability to lay down a huge volume of fire without turning your gun into a glowing mass of molten aluminum and steel. Essentially, it's a civilian version of the M27 Infantry Automatic Rifle. You can then add the belt-fed components once you've found a buyer for your kidney.

Being the kind of sneaky, underhand, "damn the man" individuals that we are, we wanted to check out some alternative options to a hugely expensive transferable AR lower or registered drop in auto sear. We ruled out the popular sliding bump-fire stock because we thought the reciprocating motion might cause belt-feeding issues. Slapping on a SIG Brace was a no-go, as Ares Defense only ships 12-inch barrels to people with a registered SBR; however, that doesn't stop you from having a gunsmith cut, crown, and thread one. We decided to take one of our lowers and install the buffer, spring, and bolt catch that ship with the MCR, along with the new Tac-Con trigger. Once we found the sweet spot of the Tac-Con's assisted reset function, we were able to lay down fire at a rate between 400 and 450 rounds per minute — about the same as a BREN gun.

If we were to sum the MCR up in one word, it would be "options." It's a true weapons system that allows an individual shooter, military, or law enforcement unit to adapt their weapons to anticipated threats. You can change them quickly in the field by swapping barrels and top covers or by simply availing yourself of the option to select modes of fire. If you have a registered full-auto lower, you definitely need one of these. If you have 10 ARs in your safe, sell half of them, buy an MCR, and start linking up 5.56mm. **R**

ARES DEFENSE MCR

CALIBER: 5.56mm

MAGAZINE CAPACITY: 200-round boxed belt or 30-round magazines

BARREL LENGTH: 12.5 and 16 inches

OVERALL LENGTH: 34.5 inches with 16-inch barrel

WEIGHT: 9.1 pounds as tested

MSRP: $5,000

URL: www.aresdefense.com

NO TRAP DOORS

Phoenix Weaponry Puts a New Spin on a Frontier Classic with the .45-70 Auto

By Tom Marshall
Photos by AZ Photoman and Steven Kuo

It wouldn't be much of an irrational argument to claim that the AR-15 platform has been refined nearly to its engineering apex. Different manufacturers have made alterations to barrel length, gas system length, buffer weight, buffer spring weight, gas port size, chamber dimensions, and just about any other physical dimension you can name. With the sheer volume of technical data now available on the rifle's performance in various configurations, America's Rifle has now become a prime test bed for all manner of ballistic experiments. In a quest to push the performance envelope of the platform, and to sell more rifles, many companies are fielding AR-pattern rifles in more and more unconventional calibers for specialty applications.

Enter Phoenix Weaponry in Colorado. They've undertaken a number of outside-the-box projects and are bringing a particularly unique big-bore AR to market. What started as a pet project will now be available for purchase. They've built a highly customized AR rifle around an equally custom cartridge: the .45-70 Auto. Phoenix Weaponry let us play with their preproduction test rifle during a group hog hunting trip in Texas. The rifle's performance against the Lone Star State's four-legged vermin didn't disappoint. Before we talk about that, a little history on where this hard-hitting round came from.

WHAT'S OLD IS NEW AGAIN

The renaissance of American small arms development that occurred during and after the Civil War gave birth to a number of innovations in the realm of shoulder-fired ballistics. Some of these wound up relegated to

the pages of firearms oddity. Others turned out to be true milestones in the history of firepower. Most fell somewhere in between. The .45-70 cartridge was developed by the U.S. Army's Springfield Armory specifically for use in their 1873 Trap Door rifle. The original full nomenclature for the cartridge, 45-70-405, was derived from the load parameters — a .45-caliber slug, sitting atop 70 grains of black powder, weighing 405 grains.

The cartridge was a replacement for the earlier .50-70 cartridge that was, from its outset, a temporary solution for Army requirements that saw less than a decade of service in the ranks. The smaller .45-caliber cartridge was flatter shooting with a longer range compared to its bigger, short-lived brother. Note that "flatter shooting" is used with a large dash of relatively speaking. The combination of a heavy slug and slow-burning black powder gave the .45-70 a rainbow trajectory.

While making accurate hits at long range was possible, and did occur with some regularity, this was heavily dependent on proper range estimation and sight adjustment, as the original loadings could generate as much as 50 inches of bullet drop by the 400-yard line. Minimum acceptable accuracy standards for the .45-70 set forth by Springfield were a 4-inch group at 100 yards, but it's said that a more accomplished marksman could hit a 6x6-foot square out to 600 yards — a technique mainly used in mass volley fire, as the effective range on point targets was capped at 300 yards. Original loadings pushed the aforementioned 405-grain bullet out of the muzzle around 1,300 feet per second.

The .45-70 saw service across the western frontier throughout the entirety of the Indian Wars. In this capacity it gained an excellent reputation against both enemy forces and large game. Its military service continued into the Spanish-American War, when the Army finally realized that its soldiers were severely outgunned against an enemy with bolt-action repeating rifles. The U.S. Army had previously chosen the Trapdoor Springfield over available repeating designs, which the brass deemed wasteful of ammunition and encouraging inaccurate fire. An interesting piece of military trivia — the .45-70 can still be found aboard U.S. Navy ships, in the form of blank cartridges used for line-throwing guns.

As a result of its successful performance on the plains, the .45-70 gained popularity among sportsmen and hunters of the time, with some medium and big game hunters using it even today. The almost-one-ounce lead pill has proven effective against traditionally difficult game, including elk, buffalo, and grizzly bear. Because of this, several companies produce modern ammunition and repeating rifles for the round. Lever-action carbines have been a staple guide gun among Alaskan bush hunters and others who routinely encounter potentially aggressive large animals.

Since the original .45-70 cartridge sported a thick rim for breach loading rifles, Phoenix Weaponry had some work to do in order to fit the round into a box-fed autoloader. What they wound up doing was to use a lathe to turn down the rim on genuine .45-70 brass. The result is a rimless case that'll fit on an AR-10 bolt face. While no commercially loaded ammunition is currently available, you have two options to roll your own. If you already have a lathe to modify

original .45-70 brass, Phoenix will sell you the form cutter. Or you can simply purchase pre-modified brass directly from them.

Since we didn't have any loading materials on hand, they pre-loaded some test ammo for us. These rounds used a 325-grain Hornady FTX slug atop 53.5 grains of IMR4198 powder. For reference, similar .458 SOCOM loadings use only 36 grains of the same fuel. Resulting muzzle velocities out of an 18-inch barrel were in the neighborhood of 2,500 feet per second. The crew at Phoenix has several other loads either already dialed in or in the works, including a 250 grain FTX bullet traveling at 2,700 feet per second and a 500-grain DGX slug that, at time of writing, had yet to be chronographed.

THE LAUNCHER

While the .45-70 Auto was designed to run off an AR-10 bolt, and the rifle looks like an AR-10 from the outside, much of the rifle

From left to right: .45-70 Auto, .308, and .223.

As with most AR rifles, the Phoenix Weaponry .45-70 Auto is pretty adept in improvised shooting positions, but at nearly 10 pounds, this isn't necessarily the first AR we'd choose to carry around all day.

is produced in house. The receiver set, dust cover, handguard, trigger, and muzzle brake are all produced by Phoenix Weaponry. The muzzle brake is what they call their chevron brake. We found it to be highly effective and highly loud. Noise, flash, and fireball are all exactly what you'd expect from sending a ¾-ounce slug down range at Mach 2.

The trigger on the test-bed rifle tripped at just over 3 pounds with a touch of pre-travel before a crisp break. The handguard is also made in-house. Rail sections, sling mounts, and hand stops can all be mounted to it. Having said that, we're not sure why it wouldn't have been simpler (and more convenient for the customer) to make an MLOK pattern tube. It's 15 inches long, shown here on an 18-inch barreled gun. Speaking of barrels, it's interesting to note that this rifle runs a 1:14 twist barrel, while original .45-70 rifles typically used a 1:20 twist — logical, considering the increased velocities out of the autoloader. The barrels themselves are 4140 CMV and button rifled.

Phoenix uses a mid-length gas system on the 18-inch rifles. If you request a 20-inch or longer barrel, it'll be a rifle length system. The BCG, buffer, and buffer spring are all standard AR-10 parts, which means owners will be able to tune their recoil systems to a specific load. Additionally, the charging handle, furniture, and fire controls are off-the-shelf as well. Feel free to add on your favorite bolt catch, mag release, or selector lever.

Magazines start life as standard AR-10 boxes, but must be heavily modified by Phoenix Weaponry to actually feed the .45-70 Auto round. We asked about the modifications, but they're playing that process tight to the vest. Once complete, you'll get 6 rounds of .45-70 into a 10 round PMAG. We're waiting patiently for them to perfect drum mags. Our test gun was cerakoted in FDE, but since these weapons will be made to order, you'll be able to request the color or camo pattern of your choice.

For a weapon of its size, our test gun was surprisingly easy to control.

Our baptismal experience with the .45-70 Auto cartridge was at Quintanilla Ranch in Tilden, Texas. Lesson one: It's not ideal on covered firing lines.

ROUNDS DOWNRANGE

We had the opportunity to take this gun out into the (hog) killing fields of west Texas during a team building hunt with Escondido Hunting at the scenic Quintanilla Ranch in Tilden, Texas. If you've never heard of Tilden ... don't worry ... neither has anyone else. But the food was good, the people were great, and it let us witness first-hand the potency of 19th century ballistic genius. On the flat range, we were zeroed within five rounds. Our gun came with a Burris MTAC 1.5-6x42mm already mounted, which made things pretty easy at 100 yards and in. We also learned very quickly that this is not the gun you want to be shooting under a covered shooting position with people

Despite some custom dimensions and unique engineering, this rifle is still an AR at heart. All the controls will be intuitive for anyone who's served in uniform or played first-person shooter games.

on either side of you — unless you hate those people.

Recoil wasn't nearly as stout as expected. The monster muzzle brake, softer-shooting mid-length gas system, and buffer/spring combination brought it down to a level that we would compare to an inertia-driven 12-gauge. Follow-up shots were definitely possible, though had to be executed judiciously.

We did get lucky enough to take a single small hog with this rifle. It was after sunset, and we didn't have night vision. After hours of no activity, the author and RECOIL OFFGRID web editor Patrick McCarthy were getting ready to pack it up for the night when they heard movement outside the blind. The shot was made through the MTAC's illuminated reticle with Patrick using his EDC flashlight to illuminate the trio of hogs approximately 60 yards from the blind.

The shot was far from ideal, catching a 40-ish-pound pig in the hindquarters. A smaller gun would've left the animal suffering. But at twice the speed of sound and then some, the 325-grain flex tip slug ended the ordeal in a matter of seconds — inverting the pig's entire back end in the process. When the two of us, our guide, and Iain inspected the crime scene, the results were impressive to the point of being almost comical. Suffice to say that .45-70 Auto can be loaded to handle almost anything in the lower 48.

The only limiting factor is price point. This rifle is listed on Phoenix Weaponry's website at $4,800, a price tag that hits almost as hard as the .45-70 itself. We concede that there's a lot of custom tuning that goes into getting these guns to function properly. But a price point that's halfway to five figures will make this gun a big-bore pipe dream for many. Those with the means, however, will get reasonably close to aircraft-grade firepower and be hard pressed to find themselves underpowered in any situation. R

PHOENIX WEAPONRY CHRISTINE

CALIBER: .45-70 Auto

MAGAZINE CAPACITY: 6 rounds

BARREL LENGTH: 18 inches

OVERALL LENGTH: 41 inches

WEIGHT: 9.75 pounds

MSRP: $4,800

URL: www.phoenixweaponry.com

FEATURED ACCESSORIES:
Burris MTAC 1.5-6x42mm - $479

PRICE AS FEATURED: $5,279

THE HEAVY-HITTING HALF INCH

Big Horn Armory's 500 Auto Max — the Most Powerful Handgun Cartridge in the World — in a Rifle.

By Tom Marshall
Photos by Chris Heising

The AR-15 rifle ... the actual AR-15 rifle, designed by Eugene Stoner and produced by ArmaLite ... was part of a U.S. Army request for a Small Caliber, High Velocity (SCHV) cartridge. A number of major ammunition manufacturers put forth candidates for the SCHV program, many based on the .222 Remington. The prevailing design became known as the .223 Remington, and it was decided that 55 grains was the ideal bullet weight for this caliber. For reference, it takes approximately 437 grains to equal 1 ounce — meaning your average 55-grain FMJ slug is just over one-tenth of 1 ounce. Compare that to the first time we fired the eyebrow-raising AR variant produced by Big Horn Armory in Wyoming, which sent a 440-grain cast lead slug down range at roughly 1,900 feet per second. That's 1.1 ounces of lead flying at just under 1,300 mph.

THE STONE

If nothing else, this bit of ballistic dichotomy illustrates the utter versatility of the AR platform. While the .222/.223 has long been a staple cartridge for varmint and small game hunters, its diminutive size had placed a thick glass ceiling over the AR's use in game-getting roles. Enter Big Horn Armory. Based out of Wyoming, Big Horn's original purpose in life was to create a Browning-pattern lever gun chambered

Even with the "in line" brake and heavy-duty honeycomb recoil pad, the .500 Auto Max still offers substantial push-back.

in .500 S&W Magnum. Why zero in on this gratuitous caliber? Despite its air of flagrant "because-why-not" amount of energy, the .500 S&W Magnum has excellent potential for big game hunters.

With bullet weights ranging from 265 to 700 grains, and several loads pushing a 500-plus-grain slug to supersonic speed, the .500 S&W Magnum is capable of dropping any and every large game animal on the North American continent. Additionally, the big 500 brings this power to states with restrictive hunting regulations, some of which only permit hunting with straight-walled cartridges. Big Horn's resulting lever gun, dubbed the Model 89, is one of the few long guns to chamber this massive magnum.

But Big Horn Armory wasn't content to rest on their laurels. They went one step further and set about to put all that power into, arguably, the most popular semi-auto rifle in the country. Except for one not-insignificant problem. The .500 Smith & Wesson Magnum, purpose built for use in their ginormous revolvers, sports a semi-rimmed case that refuses to feed from box magazines.

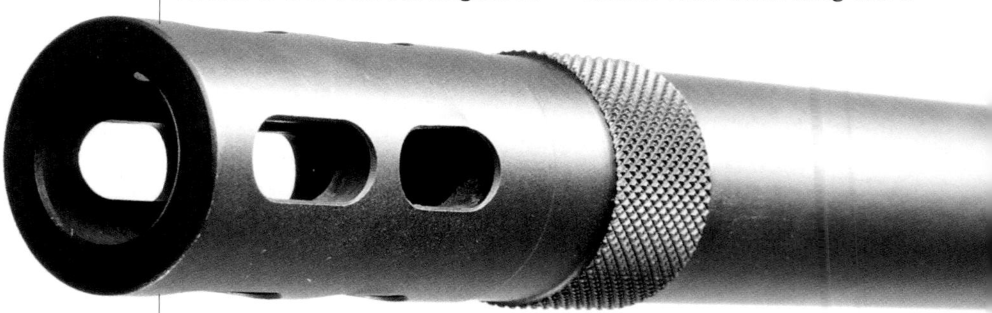

Their solution took the form of a variation they call the .500 Auto Max. This slap-your-forehead-simple fix consists of taking the original .500 S&W Magnum and turning down the rim, allowing it to cycle in semi-auto actions.

None of the ballistic specifications change, and the only limits to rebuilding .500 S&W Magnum loads in .500 Auto Max are chamber length and pressure tolerances. When compared to their lever-action offering, their AR500 rifle offers faster follow-up shots, easier reloads, and all the bells and whistles of the AR aftermarket. Think about being able to sling all that lead from a rifle that'll readily accept thermal optics, IR lasers, and a rail-mounted chainsaw. (The latter would conveniently allow you to drop both an Elk and a couple of trees for firewood.)

About this time last year, we tested another big-bore AR in another "auto-fied" cartridge, the .45-70 Auto. Similar to the .500 Auto Max, the .45-70 Auto took a rimmed cartridge and turned it into a rimless one. Whatever your opinion of the weight and velocity comparison, the .500 Auto Max enters the ring with a key advantage: factory ammo. Both Underwood and Buffalo Bore currently stock a variety of factory loadings in .500 Auto Max. Underwood offers a 350-grain JHP, as well as their solid copper Xtreme Penetrator slug in both 350- and 420-grain variants, plus a 440-grain cast lead load. Buffalo Bore has 350-grain JHP, 400-grain JFN (jacketed flat nose), and 440-grain cast lead. For both companies, the 350-grain JHP loads are marked at 2,300 feet per second, with the 440-grain cast lead listed at or near the aforementioned 1,900 feet per second. If you do prefer to roll your own, all of the components to do so are also readily available.

THE CATAPULT

The rifle itself is built on an AR-10 receiver set. The .500 Auto Max round can be stuffed into an AR-15 magazine, and the AR500's magwell is shortened to AR-15 length to accommodate the shorter mags. But building on an AR-10 base platform allows the use of heavier buffers and stiffer springs to mitigate the not-insignificant recoil. Specifically, Big Horn Armory uses an 18-pound recoil spring and a 3.8-ounce buffer. This, combined with a mid-length gas system and an "in-line" muzzle brake makes the AR500 … manageable. Felt recoil is still rather stout, maybe on par with shooting 3-inch magnum shells out of a 12-gauge pump gun, but not joint-busting.

The whole gun seems to track like an artillery piece, with a slow arcing movement to the rear with each trigger press. But the weight of the gun does bring it back down to level as part of its natural path of travel. The handguard is modular, but not to M-LOK or KeyMod standards. It uses an older-style system of screw-in short rail sections. While not quite as convenient as the current iterations of plug-and-play handguards, it does allow for user configuration. Our test gun shipped with a single 20-round AR magazine modified to accept five of the larger .500 cartridges. On that note, it'd be interesting to see how many could squeeze into a 30- or 40-round AR mag.

We topped our test gun with a Nikon BlackForce 1000 1-4x in a Warne mount. Since our first outing with this rifle was to be on a veteran's charity hog hunt, we thought an LPVO would give us a good balance of magnification and rapid engagement ability. Unfortunately, the scope doesn't smell like feed corn and, thus, didn't attract any hogs for us to collect first-hand soft tissue ballistics on the .500 Auto Max. Uncooperative vermin notwithstanding, the AR500 rifle brings all the familiarity you'd expect from a standard-form black rifle, with significantly bigger ballistics.

Nikon's BlackForce 1000 scope provides 1-4x magnification and an accessible reticle, ideal for the .500 Auto Max's short-to-medium-range trajectory.

ROUNDS DOWNRANGE

We made several range trips with the AR500 rifle after our empty-handed hog hunt. Taking it as close in as 25 yards, we got an entire magazine through one ragged hole while standing and unsupported. But don't expect to do that with any kind of record split times, as this gun will start to climb away under rapid fire. It should also be noted that the .500 Auto Max isn't a long-range caliber. According to ballistic charts on Buffalo Bore's website, with a 100-yard zero the flat nose rounds we were shooting reach their apex around 75 yards. By 200 yards, they drop just over 8 inches.

We did experience some feeding issues with the AR500. Toward the end of our testing, we started to experience increasing numbers of double-feeds. This was not an issue for us in the beginning. Upon examination of the included magazine, we noticed the feel lips were spread noticeably. Not necessarily a surprise, as the .500 Auto Max case is so much wider than that of a

The AR500 is based off of an AR-10 sized receiver set, explaining this immense, and handsomely sculpted, magazine well.

.223 — but it's a poignant reminder that, as with any firearm, magazines should be considered consumable components and ample replacements kept on hand.

All in all, the AR500 does what it claims to — it delivers massive amounts of energy against heavy targets at short range. In this role, it offers several advantages over more traditional lever guns in similar big-bore chamberings, while being significantly less expensive than most of the other big-bore ARs we've come across. If you live in a region populated by large game, or you like to demolish cinder blocks and junker cars on the range, then the Big Horn Armory AR500 is a reasonably priced demolition tool that may be just the thing to take down your first bear, or small family sedan. **R**

Premium accents like enhanced controls, ambidextrous safety and bolt catch, and multiple QD attachment points make the AR500 a strong competitor at its price point.

BIG HORN ARMORY AR500

CALIBER: .500 Auto Max

MAGAZINE CAPACITY: 5 rounds

BARREL LENGTH: 18 inches

OVERALL LENGTH: 41 inches

WEIGHT: 9 pounds, 8 ounces

MSRP: $1,999

URL: www.bighornarmory.com

FEATURED ACCESSORIES:
Nikon BlackForce 1000 Scope - $400
Warne Gen 2 MSR Scope Mount - $136

PRICE AS FEATURED: $2,535

BLACK RIFLES THAT BLEED RED

A Pair of ARs for AK lovers

by Tom Marshall
Photos by Kenda Lenseigne

The AR versus the AK. Just seeing that phrase conjures images of 1,000-page-long forum threads and backyard YouTube videos, as each party tries to prove why their chosen rifle is the one gun to rule them all. Luckily for us, and despite what your elected officials may try to tell you, America is still a nation of compromise. For those die-hard AR fans who are willing to try a mile in the other guy's shoes, two companies now offer black rifles that not only fire the 7.62x39mm cartridge, but run it out of traditional AK-pattern magazines. Before we get down to the brass tacks, it's worth considering one question…why?

"Because." In our opinion, this is a perfectly legitimate answer when it comes to guns. It's popular these days to demand that every new gun a company releases, or every purchase a shooter considers, be justified by a specific need or purpose. If the product can't be assigned to a deliberate task, it's somehow less legitimate. But just like big trucks, fast cars, and old wine, sometimes we want it just because it's new, old, rare, or just plain interesting. If one person is willing to sell it and another is willing to buy it, why get wrapped up in the details? God bless the free-market economy.

All soapbox grandstanding aside, an AR that fires a Russian round out of Russian magazines does offer some unique practical advantages. The AK-47 — and its magazines — are widely popular in many of the regions where American and Allied forces are currently engaged. An AR-pattern rifle, on which many NATO soldiers are already trained up, that can be fed local ammo from magazines picked up off the battlefield has some promise. Especially when one considers that AR-pattern magazines adapted to the Commie cartridge almost universally suck to one degree or another.

Note size and placement of magazine catches on the Rock River Arms (left) and the CMMG (right). The Mutant's is considerably easier to operate.

For those who hunt with ARs, the 7.62x39mm offers some ballistic advantage over the 5.56mm. Despite a somewhat lazy trajectory and exaggerated bullet drop, it delivers more energy and better penetration on medium game at shorter distances. In a defensive rifle, these same properties give the 7.62x39mm an advantage against intermediate barriers. For competitors and plinkers, AK ammo is almost always cheaper and just as widely available as the 5.56mm, allowing more aggressive training and practice regimens.

Whatever your want or need, there are two currently available options from two respected companies. We lined up samples from both to evaluate the pros and cons of each rifle.

THE VETERAN

Rock River Arms released their LAR-47 about two years ago. It was not the first 7.62x39mm AR-style rifle. But, to our knowledge, all previous iterations relied on proprietary AR-15-style magazines that had been resized for the shorter, thicker, and tapered case. The normally shallow curve of Stoner's G.I. magazine had to be steepened unnaturally. They functioned — just barely — but were as reliable as your deadbeat brother-in-law. Also, these magazines were (and are) tough to come by and more expensive when you could get them.

Rock River skirted the feeding problem by housing the ammunition in its natural habitat — and instead redesigning the rifle around the mags. While we don't know what kind of sales success the LAR-47 has had to date, we haven't seen or heard anything poor written about them. It has sat comfortably for the last two years as a sort of only-in-class product offering niche capability with no direct competition. Until now.

THE CONTENDER

At the time of writing, the CMMG Mk47 Mutant is less than three weeks old on the open market. CMMG clearly took its time, watching the competition mature in the market for at least two years and, we assume, carefully considering what unique suite of features they would offer on their own model, and no doubt making some last-minute changes prior to launch. The Mutant's features are in keeping with current trends in the AR community. The forearm is KeyMod, running to the muzzle, and is topped with a compensator as opposed to an A2 birdcage. The furniture is from Magpul, and the top rail is continuous from receiver to forearm. The Mk47 would not look out of place in a three-gun cart or dressed with a bipod and glass.

INTO THE BREACH

Rather than go into this faceoff with a set of parameters or a check-

list to run down with each weapon, we decided, this time, we'd just shoot the guns and see how they felt. We ran (or tried to run) each rifle with four different magazines, four different kinds of ammunition, and two different optics.

The magazines were one each of the following: US PALM, PMAG, RRA factory, and a steel surplus mag of indeterminate provenance from some godless heathen nation.

Ammo consisted of Wolf Poly 124-grain FMJ, Silver Bear Match 122-grain FMJ, Federal 122-grain JSF, and Red Army Elite 122-grain FMJ.

The optics were an EOTech 558.A65 and a Hi-Lux CMR-AK762 1-4x. Without making this a scope review, it's worth saying something about the Hi-Lux before we go any further. Hi-Lux has been around for a long time, but never really been on this gunner's radar. We're publicly admitting right now that this was a mistake. While a lot of people turn their nose up at gear made overseas, if you want to run magnified glass on an AK-type rifle, the Hi-Lux CMR would be an excellent choice. The BDC reticle is illuminated and pre-set for the 7.62x39mm's trajectory. The turret clicks are loud, positive, and tool-less. Optical clarity is impressive — the picture is bright and the colors are true. If you purchase either of the rifles in this review, or any other 7.62x39mm platform, the CMR-AK will serve you well in a variety of roles.

We started with the Mk47 Mutant. One of the first things to strike the shooter about this weapon right out of the box is how heavy it feels. It may be a combination of barrel profile and the longer handguard, but firing this weapon from the shoulder was not particularly enjoyable. All magazines locked in place and fed well. The RRA factory mag and the surplus steel mag both had some very mild play in the magwell, front to back, and laterally. The PMAG and US PALM locked up tight, with no play at all. Considering that AK mags vary widely in their tolerances, it seems that CMMG found

The affordable ($459) Hi-Lux scope proved to be an ideal partner to the x39 round.

a happy medium when it dimensioned the magwell.

We also really like the design of its magazine release, an improved version of the traditional AK paddle. The paddle is flared on both sides so it can be hit by either trigger finger. It can also be run by the thumb or leading edge of a fresh mag, as the current AK manual-of-arms would dictate. All ammunition fed and grouped about 2 MOA. We experienced one malfunction in the first magazine fired — a brass-over-bolt that required the shooter to collapse the stock and mortar the weapon to clear it. Neither gun was lubricated prior to testing and, after clearing this jam, we did not experience any other malfunctions with the CMMG.

The compensator up front and rubber padded Magpul stock in the back made the 7.62's recoil a non-issue. What was an issue was the Mutant's trigger. If your metric is a stock AR trigger, then it's not atrocious. But were we going to put one of these in our safe, a new trigger would be mandatory. The break was crisp...when we eventually found it. But there was a ton of stacking, and it was hardly consistent. On the range, we spent long pauses waiting for the break and the grouping exercise was a fairly intensive study in trigger prep.

The Rock River features a 7-inch quad rail with exposed gas block — a pet peeve of ours, but no effect on function. It's not quite as easy on the hands as a KeyMod piece, but its shorter length definitely gave the LAR-47 a better sense of balance. The furniture is Rock River OEM. The LAR magwell is cut on a steep slant, creating a sort of natural funnel and loads and unloads were noticeably quicker than with the CMMG. Unfortunately, this limits your magazine selection. The RRA factory mag seated perfectly, as did the surplus steel mag. The US PALM magazine would not seat at all. The PMAG seated, but would double-feed every time we tried to chamber a round.

After several stoppages, we finally broke down, lubricated the LAR-47 and tried the PMAG again with the same result. After several attempts, the Magpul mag was benched for the remainder of the LAR-47's testing. In contrast to the Mutant, the LAR-47's magazine release is an upside down T-shaped paddle located inside the trigger guard. In theory, this makes for easy access by either trigger finger. In practice, the latch on our test gun was so stiff that it made trigger fingers sore after just a few mag changes, and we wound up just using the support-side thumb, using a fresh mag to dump the empty is impossible due to the design.

The trigger is a two-stage Rock River National Match trigger — quite possibly the nicest off-the-shelf trigger we've experienced on an AR. The slack is easy to manage and stops dead against a solid wall, breaking crisply and resetting quickly every single time. We hoped that this would result in tighter groups but, alas, 2 MOA was still about the best we could manage, though those results were substantially easier to achieve with the RRA match trigger than the CMMG. The barrel is topped with an A2-style flash hider, which seemed to have no effect one way or the other on recoil. The Rock River's BCG is hard chromed, making cleanup a little easier. No malfunctions of any kind were experienced while shooting the LAR-47.

SURVEY SAYS...

It's safe to say that neither one of these rifles came through unscathed. Both are going to require some compromises on your part, depending on personal preference.

If I had to take one home today, it would be the LAR-47 hands down. Magazine choices and rail space would be limited. I would also have to train on my reload technique a little harder, but the Rock River's quick handling and easy trigger meet the prime directives for any serious-use carbine. Of course, you should buy the rifle that fits you best, not this writer. And either choice will give you a modular, hard-hitting rifle that can satisfy shooters on both sides of the Iron Curtain. R

SECOND OPINION

BY IAIN HARRISON

It's not unusual for two shooters to have different outlooks on guns. Admit it, how many times have you and a buddy engaged in chest-thumping over your firearm choices? Without wishing to turn this into a participation ribbon event where no one's feelings get hurt, it always pays to have a different perspective — while Tom spends his work days at the blunt end of an M4 in hot, miserable conditions, this shooter gets to play with guns in a somewhat more civilized environment. Which is why we came to a different conclusion.

Yeah, the Mk47's trigger won't win any prizes, but the guys at CMMG have made sure their creation will last just about forever by designing it around a cut-down AR-10 bolt. The 5.56mm bolt head has major problems when it's opened up to handle the 7.62x39mm cartridge, which is why we went with a 6.8 SPC case as the basis for our 1,000-yard AR-15 on page 54. By going large, they've put more metal where it counts. We also like the extended handguard, as it fits our shooting style and allows for more options when shooting around or over barricades, as well as being more comfortable in the hand.

It's interesting to note that in both cases, the guns turned in similar groups of around 2 MOA, which is pretty decent performance for non-match type carbines and lays to rest the idea that the 7.62x39mm round is inherently inaccurate.

ROCK RIVER ARMS LAR-47 DELTA CARBINE

CALIBER: 7.62x39mm
MAGAZINE CAPACITY: 30 rounds
BARREL LENGTH: 16 inches
OVERALL LENGTH: 33 inches
WEIGHT: 6.9 pounds
MSRP: $1,545
URL: www.rockriverarms.com

FEATURED ACCESSORIES:
EOTech 558.A65 - $630
Magpul PMAG 30 AK/AKM - $30

PRICE AS FEATURED: $2,205

CMMG MK47 MUTANT AKM

CALIBER: 7.62x39mm
MAGAZINE CAPACITY: 30 rounds
BARREL LENGTH: 16.1 inches
OVERALL LENGTH: 33.5 inches
WEIGHT: 7.2 pounds
MSRP: $1,650
URL: www.cmmginc.com

FEATURED ACCESSORIES:
EOTech 558.A65 - $630
Magpul PMAG 30 AK/AKM - $30

PRICE AS FEATURED: $2,310

HIGH FIDELITY

Shoot in Sports Car Style While Paying Hybrid Fuel Costs

By Tom Marshall
Photos By AZ Photo Man

There are few pursuits in life more costly than boat ownership and equestrian competition of any sort. Somehow, precision rifle competition feels like it's a not-so-distant third place in that ranking of activities. It's not uncommon to see Precision Rifle Series competitors pulling $5,000 worth of gun and glass from hard cases and shooting $250 worth of ammo during a match weekend. Add travel costs, a weather meter, ballistic computer, shooting bags, ear pro, and on and on. All that, and we haven't even covered the cost of practicing with your 6.5 Costmoor.

Much like gas prices, ammo is a futures market. Climate change, natural disasters, military contracts, and White House press conferences can influence the price per round just as much as the price per gallon. And once it goes up, it never quite falls back down to what it used to be in the good old days. While no caliber has been left unaffected by the myriad political panics of the last decade, shooting .22 Long Rifle continues to provide a substantial savings over its centerfire brethren.

Despite a brief period of purely outrageous rimfire prices, that particular corner of the ammo market has seemed to rebound fairly well, with a quick internet search showing bulk purchases as low as five or six cents a round. As we're typing this, we're looking at 5,000 rounds of Federal Game Shok for less than $325 ... and very strongly considering a typing break to whip out the old credit card.

With prices like this, it's no wonder that rimfire rifles, and rimfire ARs in particular, are making a comeback. We rounded up a handful of them — a couple of well-established market staples and a few newcomers — and gave them a thorough wringing out.

MORE THAN JUST A PRETTY FACE

Not all AR clones are created equal. It's one thing to build a .22 that looks like an AR-15. But that doesn't mean it'll run like one. While the basic aesthetic is relatively simple and can be accomplished by dropping existing rimfire actions into a plastic chassis, switches and buttons may not be faithful to an actual AR. This may or may not be a deal breaker for you depending on your needs, but it's something to consider prior to sinking your hard-earned green.

The rimfire AR market has made great strides in recent years, and several of the selections in this guide will accept triggers, fire controls, and furniture made for full-size guns. With customization being the name of the game in today's black rifle rodeo, this one-for-one crossover not only reduces maintenance and parts replacement cost, but also allows us to create true clones of our duty, defensive, or competition guns that feel and manipulate the same while costing pennies on the dollar to practice with.

WHAT IS IT GOOD FOR?

Aside from the hours of inexpensive fun that rimfire rifles offer for the whole family, there are a couple of great practical uses for those particular .22 guns that provide duplicate looks and function of an AR platform. As previously stated, they make great trainers if you already own or use a centerfire AR. In addition to the cost savings, a rimfire clone is also helpful if you find yourself confined to indoor ranges, many of which won't allow full-power rifle cartridges at all. Since .22LR is permitted on every range we've ever seen, your opportunities for effective practice increase accordingly. A red dot, white light, and couple of spare magazines will give you the ability to train everything from low-light to reloads with all the same controls as your service, duty, or defensive carbine.

For those with children in the household, these guns are a great way to get them on the range. Some kids will inevitably want to have a gun that looks just like yours. Those interested in competitive shooting can use these guns as a soft intro to two- or three-gun action matches. With a good match trigger, the right barrel, and a solid optic, a couple of these guns might make solid starters for the burgeoning PRS Rimfire circuit.

Speaking of optics, in addition to the full market of red dot and magnified tubes available for standard ARs, there are purpose-built rimfire options as well. The ones you see here come from Riton USA — a small veteran-owned company producing affordable glass that is assembled offshore and quality-controlled by Americans before final pack/ship. We used their 2-7x rimfire scope, with their quick-detach mount. The mount is pretty standard fare with two locking levers and 30mm rings that include adapters for 1-inch tubes. They also sent us their newest micro red dot. The included mount includes three different height risers, but, if none of those suit your fancy, the hole pattern is compatible with the Aimpoint-Micro-style mount of your choice.

KRISS

DEFIANCE DMK22C

MATERIALS
All metal construction

COLORS
Alpine, Black, Dark Earth, OD Green

MSRP
$700

URL
www.kriss-usa.com

NOTES
The DMK series brings a couple of really great features to the table. It's all metal, with an alloy receiver set and handguard. It features a last round bolt-hold-open and a unique barrel adapter that allows you to swap in aftermarket Ruger 10/22 barrels, offering the end user the ability to browse a pre-existing selection of barrels and optimize the rifle for their desired application. The Defiance rifles will also accept most "big boy" AR control parts. Charging handles, trigger packs, safety selectors, and pistol grips can all be transplanted from centerfire counterparts. With this degree of parts commonality, you can pretty much clone your full-power AR for a truly effective training gun or shoot the open-division gun of your dreams for a fraction of the cost.

However, that durability and modularity will cost you. Not just in the seven-C-notes MSRP, but also in weight. The DMK22C tips the scales at 6.9 pounds unloaded. Also, while the handguard is modular, it's a system of screw-in rail sections. After all that effort they sunk into parts compatibility, they skipped entirely over M-LOK and KeyMod handguard options, which would not only have made more sense but saved at least a few ounces. This left us scratching our heads.

KRISS

DEFIANCE DMK22 LVOA

MATERIALS
All metal construction

COLORS
Black

MSRP
$800

URL
www.kriss-usa.com

NOTES
Our second sampling from KRISS is modeled after the now-defunct War Sport's LVOA carbine. For the uninitiated, the LVOA (Low Visibility Operations Application) Carbine was designed for … well, we're not sure who … but touted a unique handguard/muzzle device combo that supposedly reduced flash signature. Also, you could brace the handguard against a hanging cable or piece of paracord for a hyper cool "crossbow" stabilizing technique. Now all of these dubious features that you probably don't need are available in a plinking and varmint caliber you'd never use in any of these situations.

Having said that, the Defiance LVOA is a pretty good homage to the War Sport legacy. If you already own an LVOA, this would be a great trainer to round out your collection. It also wins our highest accolades for undeniable mall-ninja sex appeal. But once you pick up the Defiance LVOA and start shooting, reality sets in. At first, you'll be impressed with how well the nearly solid aluminum handguard and 5.56mm compensator eliminate virtually all movement when you fire it. But the value of this will wear off quickly when you realize that it handles with all the grace of a manhole cover. It was by far the heaviest gun we tested at 7.1 pounds, the overwhelming majority of which is in the forend. But if you want an operator-as-f*ck bench rest gun, something to boost your Instagram post engagement, or have a teenage son whose primary shopping criteria is video game cool points, the Defiance LVOA may just be the exact thing you ordered.

NORDIC COMPONENTS

NC-22LR

MATERIALS
All metal construction

COLORS
Black

MSRP
$775

URL
nordiccomp.com

NOTES
The biggest draw of the NC-22LR may be its match-grade, stainless steel barrel. In addition to the barrel's borderline bull profile, it's equipped with a stainless steel bolt and aluminum receiver set/handguard combo. While your Nordic Components rimfire rifle will be able to be passed down for generations, it also weighs just shy of 10 pounds — that's unloaded, no accessories. The handguard is slender and seemingly ribbed for your pleasure. Like the KRISS rifles, you can screw in short sections of Picatinny rail for accessory mounting — but alas, no direct-mount modular system. If you intend to shoot any kind of rimfire competition, this is the first rifle out of the pack we'd look at. But if you're looking for a lightweight tactical trainer, you might be better off looking elsewhere in this guide.

EDITORS PICK

SMITH & WESSON
M&P 15-22 SPORT

MATERIALS
Metal and polymer construction

COLORS
Assorted

MSRP
$449

URL
www.smith-wesson.com

NOTES
The M&P 15-22 Sport gets our all-around top pick. While sticker price is $449, you can get them used or on sale in the high $300s, making them accessible to just about anybody. The S&W was also the only sample we tested that had a true M-LOK handguard, allowing you to use all the light mounts, forward grips, sling loops and other assorted goodies that may be rattling around in your garage. It accepts many full-size AR parts, including triggers and safeties. Additionally, there are aftermarket adaptors that allow you to swap handguards.

Being of largely polymer construction, the M&P 15-22 doesn't quite impart the heft of our other sample guns. But it also tips the scale at a sprite 4.8 pounds. It's fast-handling on the clock, easy to use for junior shooters, and light enough to give you plenty of ounces to play with when bolting on accessories. Standard capacity magazines are cheap and prolific, and fit in most 30-round AR mag pouches. Because of this all-around versatility, budget-friendly market position, and our downright pleasant experience shooting it, the S&W M&P 15-22 Sport gets our top pick of this litter.

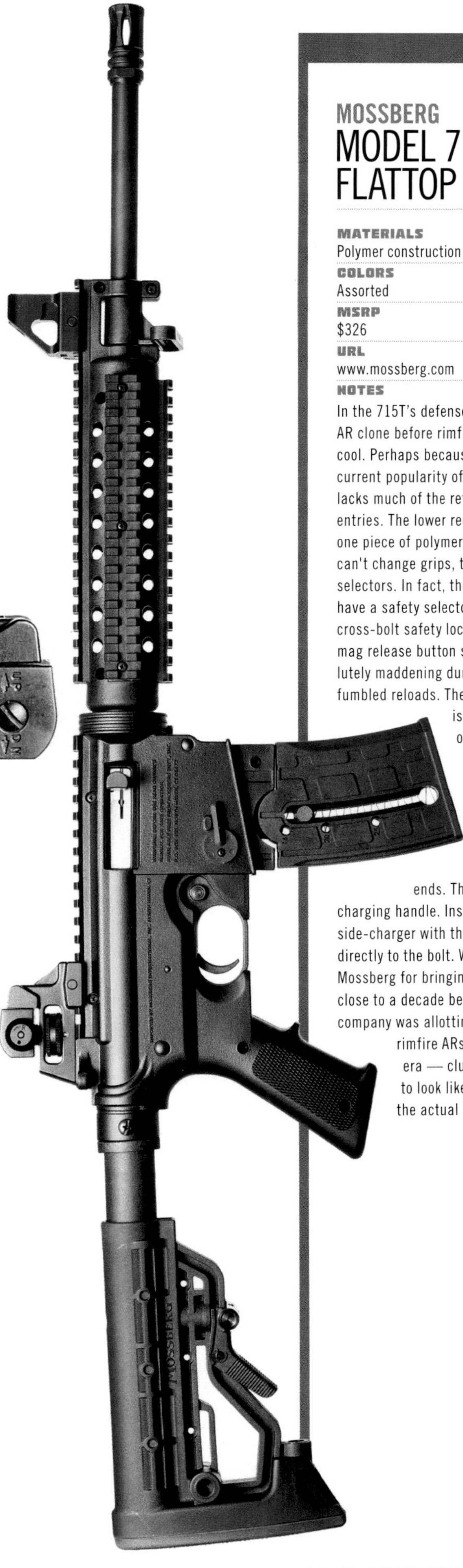

MOSSBERG
MODEL 715T FLATTOP

MATERIALS
Polymer construction

COLORS
Assorted

MSRP
$326

URL
www.mossberg.com

NOTES
In the 715T's defense, it was a rimfire AR clone before rimfire AR clones were cool. Perhaps because it pre-dates the current popularity of tactical plinkers, it lacks much of the refinement of our other entries. The lower receiver is molded as one piece of polymer, which means you can't change grips, triggers, or safety selectors. In fact, the 715T doesn't even have a safety selector. It has a traditional cross-bolt safety located right where the mag release button should be — absolutely maddening during your first 10 or 12 fumbled reloads. The actual mag release is a weird lever located on the side of the magwell.

While it does sport Picatinny rails, this is about where your modularity ends. There's not even an AR charging handle. Instead, this runs as a side-charger with the handle attached directly to the bolt. While we applaud Mossberg for bringing the 715T to market close to a decade before any other major company was allotting time and money to rimfire ARs, it's a product of its era — clunky and only designed to look like an AR without any of the actual benefits of one.

GOING

LWRCI's Army-Inspired, Piston-Driven Precision Gun

Story by Tom Marshall
photos by Straight 8

3 Precision

LONG

Way back in 2012, the U.S. Army put forth a "sources sought" notice for a Compact, Semi-Automatic Sniper System — CSASS, to save on word count. At the time, the M110 — a 7.62 gas gun produced by Knight's Armament and based on its SR-25 — was doing all kinds of good work in bad places. But it was still a little long, heavy, and generally ungainly. So the military broadcasted a request for a rifle with the same (or better) capabilities in a shorter, lighter package. A number of companies submitted samples for the eventual CSASS competition. LWRCI was one of them. This rifle is now available to civilian shooters, and we were lucky enough to get our grubby mitts on one.

Admittedly, this is not LWRCI's first foray into the AR-10-style platform. The previously released REPR also touts 7.62 knockdown power and match-grade reach. The CSASS, having been designed specifically for a military contract, has some differences. The most obvious is that the REPR has a side-charging receiver, as opposed to the CSASS's charging handle, which is traditionally T-shaped and top mounted.

Original REPRs were equipped with a four-position gas block. The CSASS is equipped with a 20-position gas block. For us knuckle-draggers, that sounds like too many positions. But this allows precision tuning for specific loads and (more importantly) specific load/suppressor combinations.

Both rifles utilize LWRCI's trademark short-stroke piston system. Both feed from SR-25/LR-25 pattern magazines. But the CSASS incorporates some product improvements over the REPR to increase accuracy potential. These include a mono-forged one-piece upper receiver, a hammer-forged chamber that (according to factory reps) allows the barrel to sit tighter against the upper receiver, and no need for a rail base.

The barrel in the CSASS is the same as the one in the REPR — a hammer-forged tube treated with LWRCI's NiCorr finish. This barrel produces good results in the REPR, but is further optimized in the CSASS by the aforementioned upgrades. The CSASS package is touted by LWRCI to be 1 MOA out of the box with match-grade ammo.

Now that we've dangled the carrot, here comes the stick: The MSRP is approximately $4,000. That's more than three times the price of our first car ... and almost twice the price of our second car ... and only a couple hundred less than our third. True story. If you choose to stop reading now, we won't hold it against you. But try taking a peek at *Going Hot* to get your blood pressure down, and then come back.

Not many of us have that kind of cash on hand. Even fewer want to take out a mortgage on a long gun. However, when you look at rifles in this class — that is, semi-auto precision rifles in 7.62 that are rated worthy of no-sh*t combat service — you won't find anything for much less. Building these guns is resource intensive, and the handful of companies who produce them on any kind of scale have to get their return-on-investment somehow. This is especially true for those brands that have chosen piston-driven designs. Conventional wisdom says piston guns face several additional hurdles to honing accuracy, such as an increase in reciprocating mass and the inability to truly free float the barrel.

What about your return-on-investment? What are you getting for four G's that you can't get for two? That's what we wanted to find out. But before we get into the nitty gritty of this particular rifle, here's a nugget of hard-earned wisdom: it takes a dedicated, knowledgeable shooter (or die-hard connoisseur) to truly quantify thousands of dollars' worth of value in a firearm. We find this to be true regardless of the type of weapon — whether high-end tack drivers, machine guns, guns with historical provenance, or even 1911s made from a crashed meteorite. (Yes, that's a real thing. Look it up.)

(Above) The CSASS breaks down just like any other AR-type rifle, making disassembly easy for just about anyone. (Below) The SureFire brake was highly effective at softening the .308 and is suppressor ready.

We know such people are out there, and we're pretty sure most of them read RECOIL, but this article isn't just for them. Whether you like to nerd out over the coolest military rifle competitions, you demand only the finest of fine-tuned lead projectors, or you're just a sucker for awesome photography, the LWRC1 CSASS is an impressive feat of human engineering that's also easy on the eyes. Now, let's take a look at what it can do.

ROUNDS DOWNRANGE

In order to put the CSASS through "group therapy," we topped it with the new SIG 3-18x FFP scope and stocked up on some high-end hole punchers. While SIG still remains relatively new to the optics game, the TANGO6 we strapped to the CSASS was great. Its locking turrets were easily adjustable, and the glass was bright and clear. It also has an adjustable illuminated reticle.

The LWRC CSASS digested all of the following loads.

- Hornady TAP FPD 110-grain FTX
- Hornady American Gunner 155-grain BTHP
- Black Hills 168-grain OTM
- Black Hills 175-grain OTM

We ran this rifle in two separate range sessions, starting with a 100-meter zero and working all the way out to successful hits at 707 meters. The rifle undoubtedly can push further, with range time being the only limiting factor. We experienced no hard malfunctions during the several-hundred-round wring out. Our only incident was a slightly sluggish bolt that failed

to go all the way into battery two times. It was toward the end of our second range session, and the weapon was clearly under-lubricated. The CSASS has no forward assist on the receiver, but we were able to push the bolt carrier into battery by jamming a thumb into the ejection port.

Admittedly, we didn't capture hard groups on paper. All our shooting in this eval was done on long-distance steel. Having said that, all our rounds were tightly clustered "center mass" on each steel gong. The company's 1 MOA guarantee seemed to hold true for us.

The spiral-fluted barrel is topped with a SureFire brake, which is really good at two things. One is reducing felt recoil. The CSASS absolutely does not buck the way we expected — or have experienced with other .308 platforms in the past. Rapid follow-up shots were a cinch. The other thing the SureFire brake is good at is kicking up a whirlwind of dust every time you pull the trigger from the prone. A couple of quick shots and you'll find yourself in a cloud of dust bound to make you look awesome on Instagram.

The first time we ran this rifle, it was on sandy gravel, and we were spitting out range grit for an hour afterward. But, in our minds, it's a small price to pay for softened recoil, fast follow-up shots, and the ability to plug our cans on right out of the box. The two-stage factory trigger breaks between 4 and 4¾ pounds. The first stage was distinct and smooth, followed by a crisp snapping break with just a little over travel. To us it was a couple ounces heavy for a great precision trigger, but keeping in mind that this was originally a military contract gun it makes total sense and performed well.

The CSASS was able to ring steel well past 500 regardless of whether we were prone, kneeling, standing, or braced off a barricade. In manhandling this rifle between positions, we became a little more aware of some of its drawbacks. With a loaded mag, TANGO6, and accompanying GDI mount, our test gun tipped the scales past 15 pounds. That's not insignificant when you're humping on a patrol, on a trail, or in Trooper Class competitions.

LWRC is one of the few companies we've seen that makes a truly ambidextrous lower, without somehow missing one of the four critical control parts.

Our other big issue with the CSASS is that it got hot. Fast. This is not the kind of rifle you mag dump, and we didn't. But even after a single magazine of quick (one shot every two to three seconds),

aimed fire, the handguard was too hot to hold. It was quickly rectified with a handy bandanna, but the problem remains. And because LWRCI's modular rail system is proprietary, you can't use any of the rail covers currently on the market to insulate your hand.

The user-adjustable, short-stroke piston system makes the CSASS highly reliable.

PARTING SHOTS

In the end, purchasing a weapon will almost always boil down to what your wallet and conscience can handle. For most of us, this author included, a rifle at the CSASS's price point is well out of reach. But that doesn't detract from LWRCI's high levels of quality and customer service, nor does it make the CSASS any less appealing in both looks and performance. If you're in the market for a fast shooting long-range rifle that looks good, runs smoothly, and delivers rounds as far as your skill allows, this long gun may be the one for you. R

LWRCI CSASS

CALIBER: 7.62x51mm NATO

MAGAZINE CAPACITY: 20 rounds

BARREL LENGTH: 20 inches

OVERALL LENGTH: 41 inches

WEIGHT: 10 pounds

MSRP: $4,000 (est.)

URL: www.LWRCI.com

FEATURED ACCESSORIES:
SIG Tango 6 - $1,800
GDI P-ROM L mount - $570

PRICE AS FEATURED: $6,370 (approx.)

SIG 716 G2 DMR

The Fat Kid Went on a Diet

In deciding on a rifle to run for the BAT Defense Aerial Interdiction course, I wanted to try something different from my previous schools and experiences. Specifically, I wanted to see how a 7.62 semi-auto rifle compared to its 5.56-cal brethren. I had quite a bit of experience shooting the latter and knew that the bump in caliber had its advantages, but most of the offerings in the big-bore semis were bulky and a bit cumbersome.

Enter SIG's 7.62 G2 DMR. The 7.62 DMR has been out for a few years, and it was built as a legitimate precision semi-auto platform. It delivered consistent sub-MOA performance from its free-floating barrel, was piston driven, and had an adjustable gas block to help tune things when a suppressor was added. One of the early criticisms focussed on its weight and overall length — it was considered heavy and bulky. Even though the rifle was still fresh off the drawing board, SIG sharpened their pencils and put the DMR on a diet.

Cutting weight was a primary design goal for the G2, as well as creating a shorter footprint. The DMRs have traditionally worn 18-inch or longer barrels, with the 16-inch variants being considered patrol models. In the Gen II, the barrel was whacked down and slimmed up, while the gas system changed to a two-position model, with the goal of maintaining accuracy while shedding ounces. The bolt carrier group went on a diet as well with lightening cuts reducing the mass wherever possible.

Most of the remaining components remained the same. There's still the continuous 1913 top rail that runs the length of the rifle. The lower receiver comes with ambi selector and magazine release, and a quick detach sling swivel is built into the forging. A nice little touch on the lower is the small rubber "wedge" that eliminates any play between the upper and lower receiver. It lends a solid overall feel to the fit of the rifle and keeps things consistent and snug. A two-stage Geissele SSA trigger delivers that ubiquitous Geissele clean break. In keeping with the ambi features, the charging handle works for both righties and lefties, and rounding out the package was SIGs new threaded taper-lock muzzle device that allows for easy use of any of their cans.

So what does all of this add up to? 8.8 pounds, 2 pounds of fat was cut from the original DMR's weight. So, how did it perform?

My first task was to simply see how bulky it felt in and around the aircraft; honestly, it really wasn't that different from any other 16-inch AR carbine. There was an increase in weight but it was actually appreciated in this application, as it kept the rifle stable while riding on the Tyler benches outside of the bird. The rotor wash can beat you up and push the rifle around, but the extra mass kept the rifle tracking more consistently. SIG's muzzle device helped for sure, but the rifle was tame to shoot. The G2 DMR ran without missing a beat, and of the several ammunition types tested, not a single reliability issue arose. Accuracy was as expected and sub-MOA groups were achievable with match ammo. R

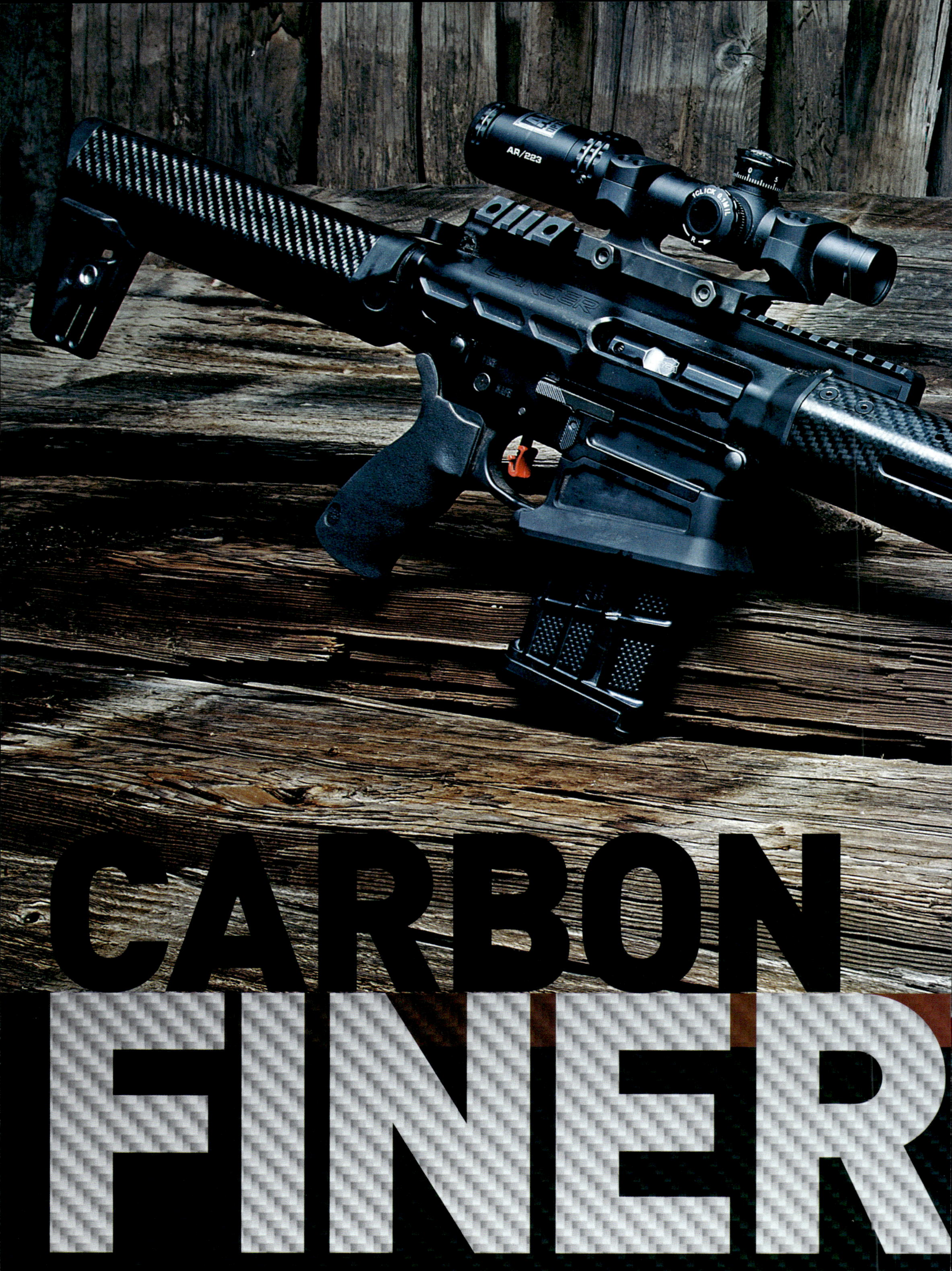

CARBON FINER

By Erik Lund
Photography by Henry Z. De Kuyper

Lancer Builds a Race-Ready .308-Caliber AR

Lancer Systems got its start making seals, way back when killing Nazis was the national pastime. No, not those SEALs — we mean the rubber kind. So what happens when a company staffed by gun nuts steps way outside its lane of expertise and tries to make a modern .308 AR rifle? Awesomeness happens.

Lancer first came to the fore with the introduction of their L5 translucent magazine. The L5 features an injection-molded, translucent polymer magazine body with steel feed lips embedded into the top portion of the tube. This steel and polymer hybrid design offers the strength of steel magazines mated with the durability and significant weight savings of polymer, all in a translucent package that allows one to visually inspect the number of rounds remaining.

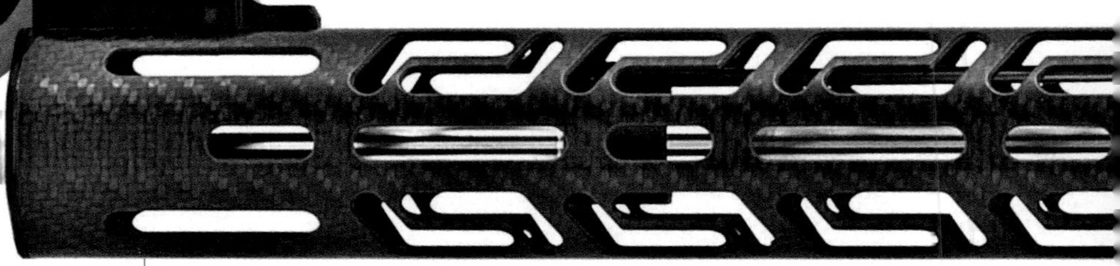

As a testament to the company's desire to continually refine and improve existing designs, Lancer recently introduced its L5 Advanced Warfighter Magazine, which now incorporates constant curve internal design geometry, more color selections, aggressive surface texturing, and stainless steel magazine springs mated with a non-tilt follower. If a company is willing to expend this much time and effort to improve the performance of a box with a spring in it, you might wonder what they might do when taking a clean-slate approach to designing a .308 AR rifle. Well, wonder no more.

STARTING OVER

So what exactly did a company that specializes in sealant technology know about building an AR-based .308 rifle? Exactly nothing. And they wouldn't have it any other way.

There were no preconceived notions, no boundaries, and no conventional wisdoms to follow. There was absolutely nothing that told its engineers what to do or where to start. There's freedom in that kind of ignorance. The freedom to look at new technologies, ideas, and concepts and not say, "It can't be done," but rather, "Let's find a way to do it." Lancer didn't fear wandering out of its lane. On the contrary, it embraced it.

Lancer's first order of business was to develop the basic framework for what would ultimately become the L30 Modern Sporting Rifle. The L30 would be a large-caliber, precision semiautomatic rifle, suitable for any application — from the recreational shooter and hunter to the special-operations soldier, and all points in between. This diverse spectrum of applications required a new approach to the concept of modularity.

THE CONTROL CENTER

Lancer's market research showed that the most prolific .308-pattern magazines currently available were the SR-25/DPMS style. Accordingly, it designed the L30 lower receiver to accept this magazine pattern. The lower receivers are precision-machined from 7075-T6 aluminum billet with the first of many innovative features — namely, an interchangeable magazine well extension.

These are attached by means of a captured rail machined into the lower receiver, allowing the extensions to slide into position. Once attached, the extensions are locked into position by the front takedown pin. The L30's lower receiver can be reconfigured with three different-sized magazine well extensions that facilitate tactical and emergency reloads. The tactical magwell extension offers a slightly oversized opening, with just enough of a bevel around the edges to facilitate a fast reload without excessive protrusion. The DMM-R/L extension offers a relief cut along either the right or left side of the extension to facilitate magazine exchanges while in the prone position. If you've ever tried to swap mags on an AR-10 while shooting from a bipod, you'll appreciate this feature. The competition extension is a bigger version of the tactical model with a larger and longer bevel, designed mainly for competition shooters. This model offers the fastest possible reload with the tradeoff of a slightly larger profile.

Moving along to the main receiver, one immediately notices a radically different surface contour designed to increase the size of the operating controls. This innovation allows the ambidextrous bolt release and magazine release buttons to be larger and easier to manipulate, without extending them out from the receiver, unlike other aftermarket solutions.

A comfortable pistol grip from Ergo Grip provides the perfect hand placement to properly place the trigger finger on the HIPERFIRE HIPERTOUCH 24C drop-in trigger system. HIPERFIRE offers a complete line of drop-in trigger systems for AR-15 and AR-10-pattern rifles and has quickly developed a reputation for providing an outstanding single-stage trigger at a reasonable price (see RECOIL Issue 9). Several of the top three-gun competitors I personally know have switched to HIPERFIRE and rave about its performance.

Rounding out the L30's lower receiver is a Lancer-designed, rifle-length carbon-fiber hybrid stock. The stock features a large, angled surface area for a more comfortable cheek weld and an extended dogleg under the rear of the stock for the support hand to stabilize the rifle when shooting from prone. Quick-detach cups are molded into the stock for use with a sling.

AR/223
WARNE
7.62 X 51
LANCER

THE UPPER DECK

Using the same design approach as the lower receiver, Lancer immediately recognized the shortcomings of the standard AR upper receiver, promptly trashed the design, and started over. Its first design change involved moving the attachment point of the free-floated carbon-fiber hand-guard away from the barrel nut and into the upper receiver itself. The distinctive hand-guard extension is machined directly into the upper receiver, performing several important tasks. This design reduces weight by allowing the use of a smaller and lighter barrel nut, but more importantly, it assures the top rail of the handguard is perfectly aligned with the upper receiver rail — no more canted or misaligned free-float tubes.

Additionally, the end user can select from several different carbon-fiber tube lengths with heat dissipating slots to customize the rifle to their needs. Carbon-fiber provides a rigid, lightweight alternative to aluminum that is stiffer and dissipates heat quicker — important during long courses of fire. Those who favor using suppressors when shooting or who shoot large volumes of ammunition very quickly will appreciate its heat-resistant qualities. Firing the same number of rounds using other handguards would cause them to quickly heat up to the point of being untouchable without the aid of gloves.

Attached to the receiver at the 12 o'clock position is another innovation, an interchangeable upper rail. The end user has the option of two differently angled 1913 Mil-spec Picatinny rails that bolt up rigidly to the upper — a standard flat rail and one machined with a 20 MOA taper for long-range shooting. Whichever you select, you'll then have to decide between three different lengths: a standard-length model that ends flush with the upper receiver, a slightly longer version that extends a few inches past the upper receiver, and a full-length top rail that runs to the end of the hand-guard.

The heart and soul of the L30 is an 18-inch stainless-steel, 1-in-10-twist match-grade barrel that sports a perfectly crowned ⅝x24-threaded muzzle and features Lancer's adjustable three-port Nitrous muzzle brake. Barrel fluting is available depending upon barrel configuration to further reduce weight.

For the next piece of the performance puzzle, Lancer turned to Rubber City Armory (RCA) for the BlacknitridePlus .308 bolt and carrier combination. RCA is an industry leader in providing high-performance, precision-machined and nitrided bolts and bolt carriers. The proprietary QPQ (quench-polish-quench) process changes the chemistry of the outer few microns of the bolt's steel, making it much harder and improving both lubricity and corrosion resistance, while reducing wear and stress on critical parts. Very few shooters in this world are harder on their equipment than pro three-gunners and a quick walk past the rifle rack at any major three-gun match will reveal a large number of RCA bolts and carriers inside the pros' guns. Completing the upper receiver is a Rainier Arms Raptor ambidextrous charging handle. Manufactured from 7075 aircraft-grade aluminum, the Raptor provides plenty of strength and leverage for working the larger and heavier .308 bolt. It features large serrated wings that are easily manipulated from either side and is a perfect match for the L30's ambidextrous controls.

TESTDRIVING AWESOME

Picking up the rifle did not disappoint, nor did Lancer's new

20-round, .308 magazine. Constructed in the same fashion as the L5 AWM, I predict this mag will quickly become the new industry standard for .308 ARs. The L30 balanced well and pointed effortlessly. The full-length stock fit perfectly in the shoulder and the carbon-fiber free-float tube filled the hand nicely with enough extra length to fit any support-hand gripping style (though, bear in mind, I'm not exactly what you'd call petite). The fluted 18-inch barrel transitioned easily between targets while the combination of the HIPERFIRE trigger and the Nitrous muzzle brake rewarded each trigger press with a loud clang on the 50- and 100-yard steel targets. Tearing through a 20-round magazine gives new meaning to the phrase "Fire at will." Hits on steel at speed were boringly easy, and it fed, fired, and ejected everything we put through it.

Before falling into our grubby paws, our test rifle had been used to burn through thousands of rounds at SHOT Show's Media Day at the Range, so we'd anticipated having to make allowances for a shot-out throat. Despite abuse at the hands of journalists who had run it like a rental car, 168-grain Federal Gold Medal Match turned in .75-MOA groups — and that accuracy could be likely tightened up with handloads. The rifle's recoil was quite soft and the muzzle tracked consistently with very little movement at all.

This rifle is a racer, pure and simple. If you're looking to scratch that .308 AR itch, then any generic rifle will suffice. But if you're looking to get into the left lane, put the hammer down, and get all fast and furious on the range, then the Lancer L30 might just be the ride you need. R

Lancer's venture in to the .308 realm points naturally, transitions swiftly, and tracks smoothly with very soft recoil. Carbon-fiber furniture looks pretty trick, too.

LANCER SYSTEMS L30

CALIBER: 7.62x51mm NATO

MAGAZINE CAPACITY: 20 rounds

BARREL LENGTH: 18 inches

OVERALL LENGTH: 40 inches

WEIGHT: 8.5 pounds

MSRP: $3,000

URL: www.lancer-systems.com

FEATURED ACCESSORIES:
Bushnell 1-4x 24mm throw down pcl - $419

PRICE AS FEATURED: $3,419

COWTOWN RANGE • MAIN EV

COWTOWN RANGE • MAIN EV

COLT VS

DMR DUSTUP

We Pit Accurate Designated Marksman Rifles from Colt and Daniel Defense Head to Head

By Iain Harrison
Photography by Kenda Lenseigne

Over the past decade, operational demands for a rifle capable of making hits beyond the range of the standard infantry weapon started popping up like weeds — from just about any major unit in contact with the enemy. The U.S. Marine Corps' 3rd Infantry Division, the Naval Surface Warfare Center's Crane Division, and the Navy SEALs, all formalized these demands into programs designed to give a squad designated marksman (SDM) the tools needed to effectively support infantry operations out to 600 meters.

Despite approaching the problem from different directions, the results were fairly similar in that all used the M16 as the basic platform, adding an aftermarket barrel, free-float handguard, and magnified optic to exploit the capabilities of Mk 262 77-grain OTM ammunition. Unsurprisingly, civilian companies were quick to jump on this bandwagon and produce their own versions of these rifles. Some, such as Knight's Armament and Daniel Defense, already had a head start due to their involvement in the original military contracts.

We decided to take a look at two examples of the breed to see what you can expect, should you decide a designated marksman rifle is in your future. Read on to find how the Colt Competition CRP20SS and Daniel Defense MK12 fare when we put them head-to-head in a RECOIL-style shootout.

COLT COMPETITION CRP20SS

Last year, when discussing matters pertaining to the Hartford company, you had to make the distinction between Colt Manufacturing and Colt Defense, who, despite sharing the same facility, were separate companies. Well, no more. Following the merger, their upstart offshoot licensee, Colt Competition was free to move into the law enforcement and military markets, so drawing on its experience in the three-gun world, it applied some of the lessons learned to produce a line of tactical models. The CRP20SS is its DMR version, and it's a beast.

UTILIZING A 20-INCH, 1:8-TWIST BUTTON-RIFLED BARREL WITH A TIGHTER .223, RATHER THAN 5.56MM, CHAMBER, THE CRP20SS IS DESIGNED TO EXTRACT EVERY BIT OF AVAILABLE PERFORMANCE FROM ITS AMMO.

Utilizing a 20-inch, 1:8-twist button-rifled barrel with a tighter .223, rather than 5.56mm, chamber, the CRP20SS is designed to extract every bit of available performance from its ammo. Colt Competition ships each rifle with a test target — this one printed a 0.371-inch group before it left the factory, so we had high expectations for it. Having used one of its 18-inch three-gun models in numerous matches, their adjustable gas system was both familiar and still unique among AR builders. One of the major drawbacks of adjustable gas blocks is their propensity to turn the rifle into a single shot at the most inopportune times. There's almost no danger of the Colt system unscrewing itself, however — in order to make any changes the user must first pull forward a spring-loaded knob, freeing it from its detent. The shooter can then control the amount of gas entering the system, so that the bolt carrier group just barely impacts the rear of the receiver extension, considerably reducing the amount of felt recoil and allowing for faster follow-up shots. A three-port compensator reduces recoil even further, to the point where the gun feels like a loud .22 LR.

A 2-inch-diameter, round-profile free-float tube provides a lot of room for accessories. When all four of the supplied rail sections are installed, the fore-end feels downright enormous in the support hand, so for most purposes we'd yank them all and cut one down to provide a small mounting section for a bipod, which would also trim some of the weight from the gun's front end. Despite tipping the scales at 10 pounds, the Colt doesn't feel too portly, mainly due to the heavy barrel being balanced by Magpul's equally chunky PRS stock. The bolt carrier group (BCG) is chrome lined, correctly staked, magnetic particle inspected, and feels a little smoother than the Daniel Defense BCG.

One of Bill Geissele's excellent Super Semi-Automatic triggers drops the hammer at 3.5 pounds and has a short, well-defined reset. Other controls are standard, save for the extended charging-handle latch which gives plenty of purchase for one-handed operation. A Magpul trigger guard and Hogue grip round out the human interface with the gun, which, like a Model T Ford, can be had in any color you like.

DANIEL DEFENSE

Although never officially recognized by Big Army, the original 3rd Infantry Division DMR rifles that were put together by the Army Marksmanship

Unit in Fort Benning, Georgia, nonetheless had a significant impact on the development of SDM programs. This is a good example of the AMU's experience in civilian competition influencing military programs and the selection of components for them, as the 12-inch, free-float handguard chosen for the weapon came from — you guessed it — Daniel Defense.

The MK12 rifle in tornado grey featured here carries the latest evolution of that handguard, and despite having rails at all four cardinal points, it's not in the least bulky. Even with the rail covers removed, the handguard isn't reminiscent of a farrier's rasp as all sharp edges have been rounded off and smoothed out. Beneath those rails lurks an 18-inch long, hammer-forged stainless barrel with a rifle-length gas system. Recognizing the limitations of chrome-lined barrels in a precision application, DD chose to nitride this one, which not only provides excellent corrosion resistance, but also reportedly increases barrel life by 30 percent over chrome linings. Barrel diameter is 0.75 inches in front of the gas block and 0.875 behind it. Unlike the Colt, its low-profile gas block is not adjustable, which shaves some weight from the front of the gun. The nitrided stainless flash hider also removes bulk, at the expense of recoil reduction.

The MK12, as with the rest of Daniel Defense's lineup, features a well-made BCG constructed to Mil-spec and chrome lined with a correctly staked gas key. Its extractor is upgraded with a black insert and donut, so tension is probably never going to be an issue, particularly considering the lower gas port pressures and shorter dwell times associated with the gas system/barrel combo. Unlike the original AMU guns, this one features extended M4 feed ramps, so getting cartridges into the barrel shouldn't be a problem either, and in the unlikely event of a stuck case, the BCM Gunfighter charging handle is strong enough to shrug off aggressive handling.

RECOGNIZING THE LIMITATIONS OF CHROME-LINED BARRELS IN A PRECISION APPLICATION, DD CHOSE TO NITRIDE THIS ONE, WHICH NOT ONLY PROVIDES EXCELLENT CORROSION RESISTANCE, BUT ALSO REPORTEDLY INCREASES BARREL LIFE BY 30 PERCENT OVER CHROME LININGS.

Both upper and lower receivers are Mil-spec 7075 forgings and on our test sample fit together with minimal slop. A nice touch is the receiver end plate, which rather than being the usual sheet steel stamping, is CNC'd from thicker material to provide a mounting point for quick-detach swivels — useful for those who use single-point slings or like to attach the back of their two-point slings at the rear of the receiver. The stock and pistol grip both include soft rubber overmolds, which in the case of the cheek-piece makes for a welcome relief from the minor annoyance of getting one's beard stuck. Should you wish to swap out the stock, a Mil-spec, six-position buffer tube will accept your preferred version. The pistol grip is more vertical and fatter than a standard A2 and incorporates an oversized, integral trigger guard which secures at the front receiver hole with a convenient Allen head set screw rather than a roll pin.

CLEARED HOT

Before we take an off-the-shelf rifle to the range for evaluation, there's usually a chunk of time spent on pulling components out of the spares bin to fix some perceived deficiency in the manufacturer's efforts.

Above: Colt's knurled gas adjustment knob tucks neatly under the handguard until needed.

Top: The MK12's pistol grip is more vertical and fatter than a standard A2 and incorporates an oversized, integral trigger guard. The stock and pistol grip both include soft rubber overmolds. Note QD sling plate at rear of lower receiver.

There's something chafing about spending valuable range time with guns that don't quite meet our (admittedly cantankerous and curmudgeonly) needs. Sort of like wearing shoes that don't fit quite properly. Both of these rifles, however, came with our favorite Geissele two-stage trigger and decent furniture already installed, so there was nothing to do apart from slapping on a suitable optic and bipod and seeing how they measured up against the clock.

Fortunately, Cowtown range had recently installed a new long-range steel array in a rocky gulley reaching all the way out to 1,100 yards, which would be an excellent facsimile of the terrain for which these rifles were designed. After familiarization and accuracy testing from sandbags using a Leupold Mark 4 3.5-10x scope (both rifles turned in sub-MOA groups with bullet weights from 55- to 77-grains) we put them into a more practical configuration.

In keeping with the original SDM concept, for our field test we used Hornady 75-grain Steel Match ammo, a Trijicon 4x32 ACOG, and an Atlas bipod, in order to engage steel silhouettes from 100 to 600 yards. Standing behind an unloaded rifle, the shooter would drop into position, make ready, and hit each target with three rounds as fast as possible, to gauge both practical accuracy and ease of manipulation. To get familiar with the course of fire, we shot it a few times with both an Aimpoint-equipped carbine and a precision bolt gun, as the DMR rifle is supposed to slot in between the capabilities of these two.

Some takeaways from the exercise:

1. Hitting an 18x24-inch steel target in natural terrain at 600 yards is a lot easier with a magnified optic. Once the paint wears off, hitting it with a red dot is a matter of luck and volume of fire, not skill.

2. It was easier to make multiple

hits with the Colt Competition CR20SS at all ranges. Its combination of heavy barrel, intrinsic accuracy, effective muzzle brake and tuned gas system meant the crosshairs never left the target and hammering steel was just a matter of managing the trigger reset.

If you're bigger and fitter than this 180-pound meat sack, or if the majority of your shooting is from prone or a solid barricade, then this rifle has all you'll need.

3. Despite the Colt turning in faster times on multiple target engagements, if you plan on humping your rifle anyplace, then the Daniel Defense MK12's 2.6-pound weight advantage is worth the tradeoff. What it gives up in stability and split times from supported positions is offset when it's deployed as an all-purpose carbine. In offhand shooting, it transitions faster and points more naturally due to its lighter and shorter barrel — overall, it's a more rounded package. R

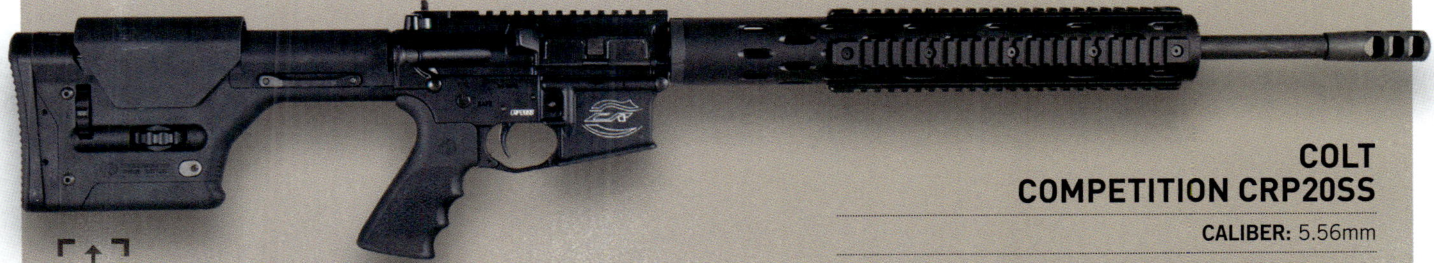

COLT COMPETITION CRP20SS

CALIBER: 5.56mm
MAGAZINE CAPACITY: 30 rounds
BARREL LENGTH: 20 inches
OVERALL LENGTH: 40 inches
WEIGHT: 10.1 pounds
MSRP: $1,990
URL: www.coltcompetitionrifle.com

FEATURED ACCESSORIES:
Leupold Mark 4 LR/T 3.5-10x40mm, M5 Front Focal - $1,875
B&T Industries Atlas Bipod (BT10-LW17) - $280
GDI P-ROM L-Model Mount - $565

PRICE AS FEATURED: $4,710

DANIEL DEFENSE MK12

CALIBER: 5.56mm
MAGAZINE CAPACITY: 30 rounds
BARREL LENGTH: 18 inches
OVERALL LENGTH: 34.5 inches
WEIGHT: 7.5 pounds
MSRP: Standard Black finish - $2,199
Tornado Cerakote finish - $2,329
URL: www.danieldefense.com

FEATURED ACCESSORIES:
Leupold Mark 4 LR/T 3.5-10x40mm, M5 Front Focal - $1,875
B&T Industries Atlas Bipod (BT10-LW17) - $280
GDI P-ROM L-Model Mount - $565

PRICE AS FEATURED: $4,919

GOOD OMEN

NEMO's Precision Rifle Is an AR on Steroids

By Sebastian Gorka
Illustration by Jenni Sisk

OK, let's clear the air first. I have issues with the AR/M-16/M-4. Don't get me wrong, I love its ergonomics and its Erector Set simplicity — my son built his first AR in his bedroom from parts in one evening at the ripe old age of 14. You just can't do that with an ACR or G-36. Call me old fashioned, but it's a caliber thing. How can I take a weapon seriously when it shoots a glorified varmint round that in many states you aren't even allowed to use on whitetails?

As an army reservist in the United Kingdom, I cut my teeth on the Brit version of the Free World's right arm, the 7.62mm L1A1. So, I'm biased. Anything under .30 is not a "battle rifle" to me, and if it won't work on deer, I don't want to use it against two-legged tangoes. For serious work, I reach for an FAL, a G3, a SCAR, or at worst, an AK-47, preferably worked over by Krebs.

But having become an American now, as part of the naturalization process it's almost a requirement to swear fealty to Eugene Stoner's masterpiece of engineering. Or at least, it should be. At the same time, we now have over a decade of data and stories from the operations in Operation Enduring Freedom and Operation Iraqi Freedom on the limitations of the 5.56 NATO round, especially at distance. Just the mere fact that ancient, Vietnam-era M-14s were scrubbed of cosmoline to be used in Iraq and Afghanistan tells you that sometimes problems can only be solved with a bigger bullet.

Subsequently, we've seen some nifty experimentation and the commercial development of new rounds that can be squeezed into the AR-15's receiver. These include the 6.5 Grendel, the 6.8 SPC, the .300 Blackout, and even the .50 Beowulf. We'll be taking a closer look at several of these in a future article that will blow away some of the myths and urban legends that are fed by the Internet. But here, however, we'll check out something even more impressive than those new and rediscovered cartridges.

SUPER AR

One of the holy grails of modern warfare is rapid and accurate fire at long range. NEMO Arms' answer to this is a super AR called the OMEN — in .300 Win Mag.

The story begins with retired U.S. Army Major General Paul Vallely and the need for a bridging platform between the dedicated .308 sniper bolt gun or SASS.308 semi-auto, and .50-caliber bruisers like the Barrett. Instead of taking Stoner's platform and squeezing more out of it, or instead resorting to yet another wildcat wonder, NEMO started with a blank slate and worked from the ground up, building a new, improved AR around a caliber that can get the job done. One that is already in the U.S. Army quartermasters supply chain, namely the .300 Win Mag. The objective throughout was to satisfy two apparently contradictory requirements: stability and recoil management to enable rapid engagements on multiple targets and real portability. Vallely and the NEMO team have certainly succeeded with the OMEN.

STRIKE SYSTEMS®
STEINER
GERMANY
1 CLICK = 0.1 mRAD
NEMO ARMS

P2 Formed from the 375 H&H Magnum, the 300 Win Mag case is wider and longer than the 7.62 NATO, which means it can drive a 175gr bullet almost 600 feet per second faster.

The OMEN platform really does look like a Super AR, and it performs like one. Since starting out in 2011, the company has introduced a full lineup of models from the lighter and shorter RECON with an 18-inch barrel, weighing in at a pretty reasonable 9.5 pounds, to the version we tested, the full-on WATCHMAN. Thanks to its 24-inch barrel and 20-MOA rail system, it tips the scales at 12.5 pounds, still not bad when you consider what it delivers.

Outwardly, the mechanism looks like an Armalite/Stoner design. The major external differences are that this rifle uses an adjustable gas-impingement system (with four settings) and the usual centerline charging handle is gone. Instead, a bolt handle is on the right hand side of the bolt carrier group (BCG), doubling as a forward assist. This design also obviates any issues that the traditional charging handle has with sniper stocks and cheek-risers. The Watchman comes standard with a Magpul PRS stock.

The OMEN's upper and lower receivers are billet 7075 aluminum, stretched to accommodate the long .300 Win Mag round. When introduced, the rifle originally utilized a 1:10 twist barrel; later versions are now fitted with very sleek 1:8 twist, fluted tubes, in order to stabilize the heaviest projectiles available. Keeping recoil in check is an effective and proprietary five-baffle brake called the A10.

To wring the most practical accuracy out of the barrel, a Geissele two-stage SSA Enhanced trigger comes standard. Clip on I2 and thermal optics are accommodated by the rifle-length free float tube, which like the upper it attaches to, is hogged out of bar stock.

However, the real "secret sauce" — and the design element that really makes the whole platform so very special — is the BCG itself. The OMEN is a response to actual battlefield needs. As a result of the access available to them, the NEMO designers were able to work with Tier One spec-op units to find a solution to the problem of how to engage multiple targets past the 600-yard envelope, and even out to a mile away.

Their solution is as effective as it is simple. To complement the effective muzzle brake, NEMO engineers added an extra spring-loaded recoil-absorbing mechanism to the end of the BCG. Instead of having just the usual synthetic-tipped buffer reciprocating in the receiver extension, the OMEN in effect has a double buffer, which really works to control impact at the end of the BCG's travel. The "keep it simple, stupid" principle really works in this case.

BOOM!

Testing long-range sniper rifles is easy only if you live in the wilds of Wyoming or the plains of Texas. When you're near the nation's capital,

"THE COMPANY GUARANTEES SUB-1 MOA AT 1,000 YARDS."

you need to seek out one of the specialized ranges in West Virginia to really see what a gun like this can do. Nevertheless, once the OMEN arrived from the NEMO factory in Montana (where else?), it was clearly in need of a functioning test, so we took it out to the NRA Range. The NRA is a short, indoor facility, so while you won't be able to demonstrate the full potential of a gun like the OMEN, you are going to seriously piss off other shooters when dropping the hammer on a .300 Windy. On that first excursion to the range, we tried heavy-grain, full-house loads from the standing unsupported position. The results were remarkable.

As I noted above, I like the 7.62mm round in all its permutations, even in superlight platforms, so

I'm fortunate enough to have a bunch of them. But the .300 Win Mag OMEN recoils less than any of my .308s — even my SCAR-17, which is a pretty soft shooting rifle. It's worth repeating — this is a .300 Win Mag that recoils less than a .308 and can be shot all day without inducing flinch. (If you don't believe me, check out online videos of small-statured women firing the lighter OMEN Recon off-hand!) The $64,000 question is now, "How long will it take for NEMO to apply the double-buffer system to other calibers for standard AR receivers?" This could be an enormous plus, especially for those super-heavy cartridges such as the .50 Beowulf or .458 SOCOM.

The OMEN's overall quality of fit and finish is right up there with any of the more established and bigger names, which is one way of justifying the gun's price tag. It's a semi-custom proposition with very little in the way of economies of scale, so you won't get much change back from seven grand. In fact, you won't get any change at all. The company guarantees sub-1 MOA at 1,000 yards for the civilian version of the OMEN and, when taxpayers are footing the bill, sub-½ MOA for their military ones. When we finally had an opportunity to put some serious glass on the beast (Steiner's Military 5-25x56), and stretch its legs a little, then it was clear that the guarantees are solid and that this gun will mechanically outshoot the operator behind it. (Check out YouTube for video of an OMEN engaging an explosive target at 1,600 yards.)

What else can one say? From the hand-anodized external finish which can be had in multiple funky designs to suit all tastes, to the nickel boron BCG and the sleek interlocking fore-end that mates with the receiver, this is one classy act. In fact, just to showcase its talents, NEMO has built a proof-of-concept .308 AR totally out of titanium, which can be yours for a cool $100,000.

The big question remains whether an average gun owner would buy one. As for me, my mom fed me well, so I generally make 1911s and M-4s look dinky. At 6 feet 3 inches, I can actually carry and use this gun fairly easily. Plus, it's a 14-shot self-loader slinging OTM bullets with a better ballistic coefficient than .338 Lapua and is able to reach out to a mile. So the answer is yes, I would buy one. People fret about the throat erosion of the .300 Win Mag, but practically speaking we're not talking about a 1919 machine gun. If you can come up with the price of admission for an OMEN, well, another barrel isn't going to break the bank. This is a precision weapon, purpose built to accomplish a specific mission that other weapons and calibers cannot.

Any complaints? One. I hate the skimpy Hogue rubber pistol grip. Yep, that's it, so sign me up. R

ABOUT THE AUTHOR

Dr. Sebastian Gorka — or Dr. G, as he's commonly called — is a freelance writer and another American with a suspiciously British accent who served in the British Army reserves. As an expert on counterinsurgency and counterterrorism based near Washington D.C., he is an irregular warfare instructor for the U.S. government.

NEMO ARMS OMEN WATCHMAN

CALIBER: .300 Win Mag
MAGAZINE CAPACITY: 14 rounds
BARREL LENGTH: 24 inches
OVERALL LENGTH: 47 inches
WEIGHT: 12.6 pounds
MSRP: $7,375.99
URL: www.nemoarms.com

PIG-POPPING PISTOL

Long Range in a Small Package with the Lantac 6ARC

By Dave Merrill

The folks at LANTAC first caught rumor of an exciting new round in development for the AR-15 back in 2018. They were part of a military research project looking for both a new rifle and corresponding intermediate caliber. In January of 2019, they were among a small, select group of companies handpicked to further develop the 6ARC and the rifle to go with it. Hornady, PROOF, and E-Lander were also a large part of this project.

Once they had their base design with Hornady supplying the ammunition, Lantac decided to invest a lot of research and development on the capabilities of shorter barrels with 6ARC. Because some companies actually do the R&D, Lantac started with an 18-inch barrel with an extremely short gas system, then cut off an inch at a time while tracking ballistic results and capabilities. Once they hit the 15-inch mark, they began chopping off a half inch at a time. In particular, we were told that 14.5 inches was phenomenal in terms of increased range compared to its 5.56 brethren.

But they didn't stop there. Everyone involved was surprised at the capabilities of the 11.5-inch barrel for its size, so they settled on that for a shorty. Thus, the impetus of this article: what we dub the Lantac 6ARC Pig/People Popper.

300BLK, 5.56, 6ARC, and 7.62N compared.

DESIGN

We should first note that we're using a preproduction prototype for this review. Some details may change prior to full production, and we'll note anticipated changes along the way. For the roots of this stubby 6ARC, Lantac started with their LA-SF15 UTP LAW Pistol and made modifications from there. You get the same receivers, SPADA-ML M-LOK handguard, LAW folder, and ambi controls; you just get a bit more to top it all off.

Our version came bundled with the single-stage, 3.5-pound E-CT1 trigger. This co-branded Timney cassette-type trigger delivers a crisp and light break, especially when using the bottom of the bow.

An 11.5-inch 1:7.5 twist PROOF barrel is the heart of this rifle. Even with this diminutive barrel length, Lantac forgoed the traditional carbine-length gas system and went with a mid-length system for a smooth impulse. This is especially impressive when you consider a mid-length gas system can be finicky to dial in even on a 14.5-inch gun, but Lantac consistently uses longer gas system lengths than tradition dictates.

Aside from the caliber itself, the most innovative component here is Lantac's enhanced bolt carrier.

Lantac has never really made a "standard" bolt carrier group; they've always incorporated at least some big upgrades from Stoner's original design. The heavier Lantac E-BCG for the 6ARC is the most advanced that they've produced to date. Starting with a billet of 8629 steel, the carrier is machined then coated with nickel boron via the diamond NiB polish process for an extremely slick and slippery result. The bolt is Carpenter 158 and magnetically particle inspected, like we'd expect from any quality manufacturer. As with all other Lantac carriers, the 6ARC E-BCG features their domed head cam pin, designed to reduce wear both within the receiver itself as well as the cam-pin track in the carrier itself.

In order to keep gas out of your face, especially when shooting suppressed, Lantac added some additional gas vent holes, some cut at an angle to blow gas forward during cycling. These also aid in keeping the bolt in battery slightly longer, simulating a longer dwell time and allowing for smoother cycling. All said and done, Lantac says their estimated bolt life is three times

Lantac never really used a "standard" BCG.

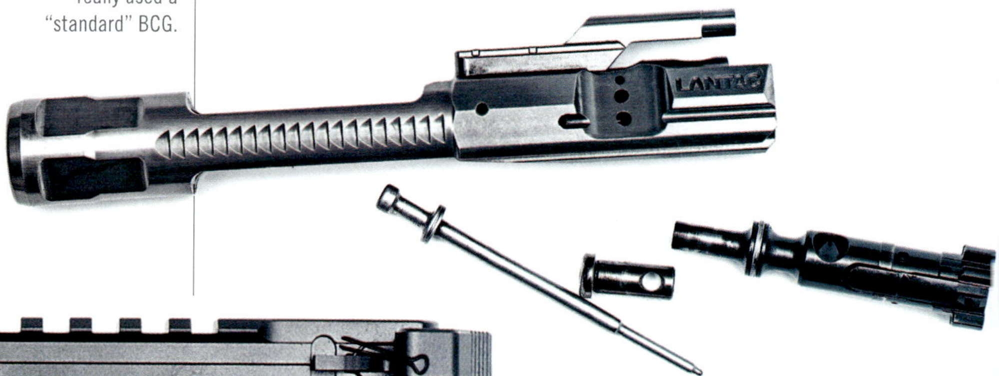

The base is a Lantac LA-SF15 UTP LAW.

longer than the current standard-issue bolt.

BALLISTICS

We can't pretend we haven't seen intermediate calibers in AR-15s before, like the 6.8SPC, 6.8SPC II, 6.5 Grendel, and more. So, it wasn't a surprise to learn that the parent case of the 6ARC stems from the 6.5 Grendel, which itself is kind of an incestuous mash-up of .220 Russian and 7.62x39. None of those rounds really caught on outside of niche groups and hog hunters, but the backing of several successful companies plus real-world use in the Department of Defense gives us confidence to say this one may break away from the wildcat-cranny into the wider world of shooting.

6ARC is initially available in 103-, 105-, and 108-grain loads. Undoubtedly, more will be released as popularity increases in both the public and private sectors.

The most effective demonstration of the ballistic capability of the 6ARC is to directly compare the Lantac 11.5 with 108-grain ELD match to the Knight's Armament 20-inch barreled M110 7.62x51mm sniper rifle shooting 175-grain SMK.

Impressively, these two vastly different rifles diverge by no more than about 0.7 mil — and that's at 1,000 yards. You'll also note that when shooting the full kilo, 6ARC has more velocity than 7.62x51mm.

WEIRDNESS

Magazines may be an issue and are undoubtedly the main issue you'll hear with initial reviews. While 6.5 Grendel AR magazines could ostensibly be used, we received a pair of dedicated 6ARC Israeli E-Lander magazines. While

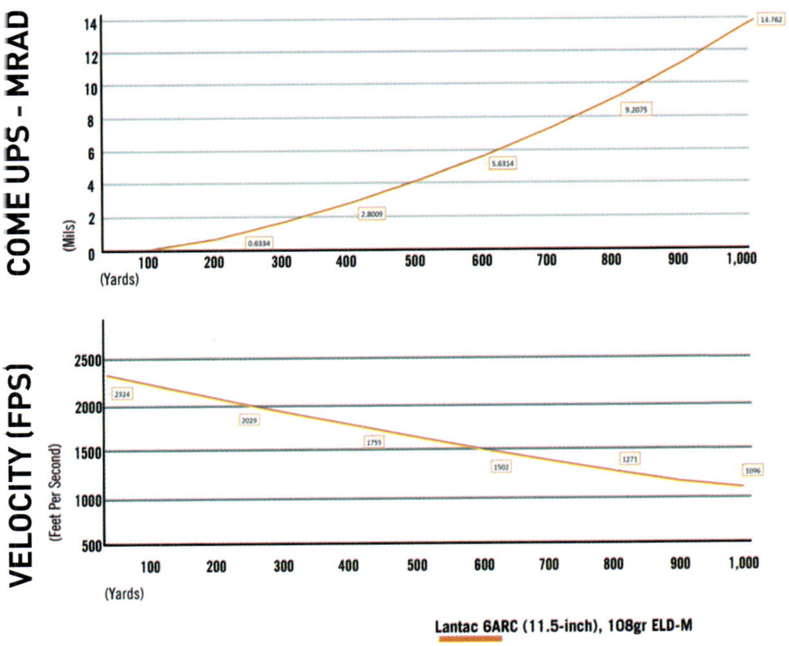

The angled gas ports on the E-BCG push gas away from the shooter.

they're about the size of a 20-round 5.56 magazine, each one holds 17 rounds of 6ARC. Similarly, what appears to be a 30-rounder only holds 24. A cursory examination gives the impression that the 6ARC magazine is no more than a 6.5 Grendel mag with new markings. Time will tell if further refinements will be needed, similar to Magpul tailoring PMags for 300BLK even though that round feeds well from standard 5.56mm magazines, for the most part.

All 6mm rounds suffer from velocity increases during barrel break-in; the 6ARC is no different. All told, you'll have to chew through around 150 rounds before these velocity increases level off, which is why we're reviewing an already-fouled preproduction barrel. Precision rifle competitor Ryan Hey from MagnetoSpeed explained this phenomenon:

"…a bullet speeds up due to the lands on the rifling being smoothed out as bullets pass over their surface. This increased bearing surface of the bullet increases pressure, thus the increase in velocity. Copper also fills in the microscopic divots in the metal, but the lands being smoothed out has a greater effect."

Therefore, you should closely track your first few hundred rounds with a new barrel with the aid of a chronograph to determine your ultimate velocities.

OUTFITTING

An easy choice for an optic, even though this is a short barrel, was the Vortex 1-10x FFP. While some would call that overkill for an 11.5-inch rifle, the 6ARC is very capable of long-range accuracy, and we're no strangers to LPVOs on SBRs. For a white light, we fitted the SureFire Scout Pro M640 and SR-07 switch with an Arcane Concerted ARCband for wire management. The M640 features an articulating mount and can be directly attached either to Picatinny or M-LOK handguards. The forward controls are an Arisaka finger stop used in conjunction with a Unity Tactical VFG (converted to M-LOK from KeyMod).

The LAW Folder allows for an extremely portable package.

Since we decided on an LPVO for an optic, we ditched the pistol lower and went with a previously registered SBR lower. All of Lantac's components were successfully transferred to this lower. We tossed the LAW Tactical folder on it but replaced the brace with a B5 Systems SOPMOD stock.

While the Lantac comes equipped with a regular .30-caliber Dragon

LANTAC USA LA-SF15 UTP LAW 6ARC

CARTRIDGE: 6MM ARC

BARREL LENGTH: 11.5 INCHES

OVERALL LENGTH: 21.5 INCHES (FOLDED), 28 INCHES (OPEN)

WEIGHT (UNLOADED): 6.4 POUNDS

MAGAZINE CAPACITY: 17, 24

MSRP: $1,855

URL: LANTAC-USA.COM

ACCESSORIES:
SUREFIRE M640 AND SR07: $411
VORTEX 1-10X FFP: $2,900
ARISAKA FINGER STOP: $28
UNITY TACTICAL VFG: $28
MAGPUL PRO BUIS: $190
B5 SOPMOD STOCK: $95
ARCBAND: $6

PRICE AS CONFIGURED: $5,513

muzzle brake, we swapped it for a Dear Air and KeyMo compatible Lantac muzzle device. Any Dead Air rifle can drops right on, as will any suppressor with a universal HUB 1.375x24 rear-end combined with a KeyMo adapter. Any silencer with 1.375 HUB mount gives you dozens of device options right from the gate.

We added a pair of Magpul MBUS Pro BUIS, because even in 2020 we still need BUIS to avoid ridicule on the internet.

Once we got our sights, optics, and controls sorted out, it was time for the range.

AT THE RANGE

Sometimes a particular combination of ammo, muzzle device, and silencer gave us good results, and other times the opposite; at first, we thought we were dealing with strange barrel harmonics. Ultimately, we found it to be a heat issue. From a cold bore, we'd get sub-MOA groups (just under 0.9 MOA), which would open up to nearly twice that once the barrel was heated over the course of two magazines. Keep in mind that this is a pre-production barrel, and the variation may not be as drastic with production models. Also remember this gun isn't intended for PRS competitions but to replace 5.56mm — even our largest groups were smaller than standard M855/A1.

We tried out several silencers, settling on the Dead Air Sandman-S with a 7.62 flash hider endcap. Not only did it consistently group well, the shift when attaching the silencer was exactly 1 mil low, making for incredibly easy adjustments when popping small targets at range.

We found it simple to engage MOA targets out to 800 yards. After it goes transonic, projectile flight becomes erratic. Still, we got hits on a 12x12-inch plate at 1,000 yards, with the main limiting factor being the center aiming point of the Vortex 1-10x reticle obscuring a target of that size.

This isn't a precision rifle capable of a combat role. This is a combat rifle capable of precision.

THIS ISN'T A PRECISION RIFLE CAPABLE OF A COMBAT ROLE. THIS IS A COMBAT RIFLE CAPABLE OF PRECISION.

LOOSE ROUNDS

If the 6ARC doesn't put the NATO 5.56 on life support, it won't be because the 5.56mm is better, but because Hornady simply doesn't currently have the machines to produce 6ARC in quantity — all the machines are running 5.56 and 9mm around the clock to keep up with the never-ending pandemic and pre-presidential panic. In every measure of capability, sans magazine capacity (which is being addressed as we type these words), the 6ARC totally beats the pants off of everything available that fits within the confines of a standard AR-15 footprint. 900 yards before transonic with a short barrel? 5.56 eat your damned heart out.

4 Lightweight

ANOREXIC AR-10

We send Armalite's Super SASS to Fat Camp

By Keith Wood
Photography by Henry Z. De Kuyper

After hearing about a few too many "game changers" that didn't live up to their marketing hype, I've become pretty jaded when it comes to new products in the firearms industry. When I sat down with a company called Proof Research at the 2013 SHOT Show in Vegas, I had my BS detector with me. They were claiming to have perfected the process of building rifle barrels out of carbon fiber. Well, technically, it's actually wrapping really thin steel barrels in the space-age compound. I'm always up for carrying less weight so I was intrigued. When they offered to put a barrel on the Armalite Super SASS languishing in my safe, I figured I'd let them prove their claims.

Proof Research is based in the emerging gunmaking Mecca of Western Montana and encompasses four previously independent companies. The firm combines a small custom rifle company specializing in ARs and bolt guns, a traditional steel barrel factory, its carbon-fiber barrel operation, and a synthetic stock manufacturer. Proof Research builds rifles from the ground up, as well as sells components to consumers and manufacturers alike. Though I've used some of its other products, which are first-rate, it is the innovation in the realm of carbon-fiber barrel construction that is making waves in the firearm industry.

Carbon fiber is roughly seven times stronger than steel by weight, and though it has been around as long as the AR-10, it didn't see widespread use on firearm parts until the past decade. Even now, the process of producing carbon fiber barrels that meet claims of strength, accuracy, and heat dissipation has proven difficult for even the best manufacturers. Using fibers traditionally found on race car brakes and wrapping them in a manner that

AAC mount serves as first baffle in the suppressor stack, reducing erosion on the can and is an efficient muzzle brake on its own.

favors the longitudinal heat dispersion of the materials, Proof Research's barrels are designed to dissipate the heat emanating from the steel barrel rather than insulate it. Proof's barrels begin as full-profile cut-rifled match grade 416R stainless steel barrel blanks made in-house that are turned down to a thin profile everywhere but the shank, muzzle, and gas block journal. This pencil-thin steel is then wrapped with carbon fiber in a helical pattern and secured with a proprietary bonding agent. The process provides a finished product that is stiffer, stronger, and 20- to 64-percent lighter than a steel barrel of equal capability — and it dissipates heat far faster, too. Stiffer barrels mean less harmonic "whip," resulting in better accuracy; the improved heat dissipation means that the barrel will maintain that accuracy over longer sessions of fire.

Today's soldiers have access to more high-quality weapons and gear than ever before, but they also have to lug it around the battlefield in challenging conditions. Lightening the soldier's load has become a major focus in defense procurement, and cutting the weight of the soldier's rifle is an ideal place to start. Proof's lightweight sniper rifle barrels have passed preliminary testing by the U.S. military and are the first carbon-fiber barrels to do so — they are currently being fielded by at least one elite unit. At this point, Proof's barrels are rated for semi-auto use only, but they are constantly innovating the product with plans to make the technology available for full auto and crew-served weapon use in the near future.

On its way from ugly duckling to swan. The stock handguard offers plenty of opportunity for weight saving.

PRI handguard is longer, which means less chance of resting the barrel on your pack and throwing the shot.

SLIMMING THE FAT SASS

We started our own testing with a box-stock Super SASS, essentially Armalite's entry into the Army's trials for what became the M110 Semi-Automatic Sniper System. The Armalite came equipped with a 20-inch match barrel in 7.62x51mm, an aluminum quad Picatinny free-floated rail system, a match two-stage trigger, and an adjustable gas block for suppressor use. The Super SASS is an excellent turnkey setup for someone looking for a semi-auto precision rifle, but at nearly 12 pounds with no optic, it's a heavy bastard. With a third of the rifle's mass coming from the heavy-profile stainless steel barrel, it was obvious that we would start there.

Since we were swapping barrels, I decided to change calibers as well. The 7.62x51mm is an amazingly versatile cartridge, but there are more suitable options available if long-range shooting is your game. While the military is stuck with 7.62mm for logistical and institutional reasons, I am not. Within the confines of the AR-10 receiver and magazine, the .260 Remington (which is a .308 Winchester case necked-down to .264 inches/6.5

mm) strikes what I believe to be the best balance between velocity and bullet mass for putting rounds on target at extended ranges. Bullets fired from the .260 Remington will drop about 10 feet less than those of a similar sectional density from a 7.62x51 at 1,000 yards. The higher velocity produced by the .260 Remington also means a shorter bullet flight time and, subsequently, less wind drift. On the other hand, the .260 isn't moving so fast that barrel life is an issue — it's "just right" for my purposes.

Proof Research built me a 22-inch carbon-wrapped barrel chambered in .260 Remington with a 1:8 twist for stabilizing high ballistic coefficient 140-grain bullets. They traded the Armalite two-position gas block for a fully adjustable unit that fit one of their standard barrel sizes, as we didn't have time to order a custom profile barrel. The full-contour steel section at the muzzle was threaded to accept an AAC Blackout to cut down on recoil and to allow for the attachment of a 762-SDN-6 suppressor. By swapping out the factory barrel and gas block for the Proof Research barrel, we immediately cut 2 pounds 2.5 ounces from the overall rifle weight, while adding 2 additional inches of barrel length. In addition to the weight savings and improved performance of the carbon barrel, it just plain looks cool: the unique subdued black and gray camo pattern is a byproduct of the proprietary manufacturing process. Everyone who sees the smooth, swirly surface can't help

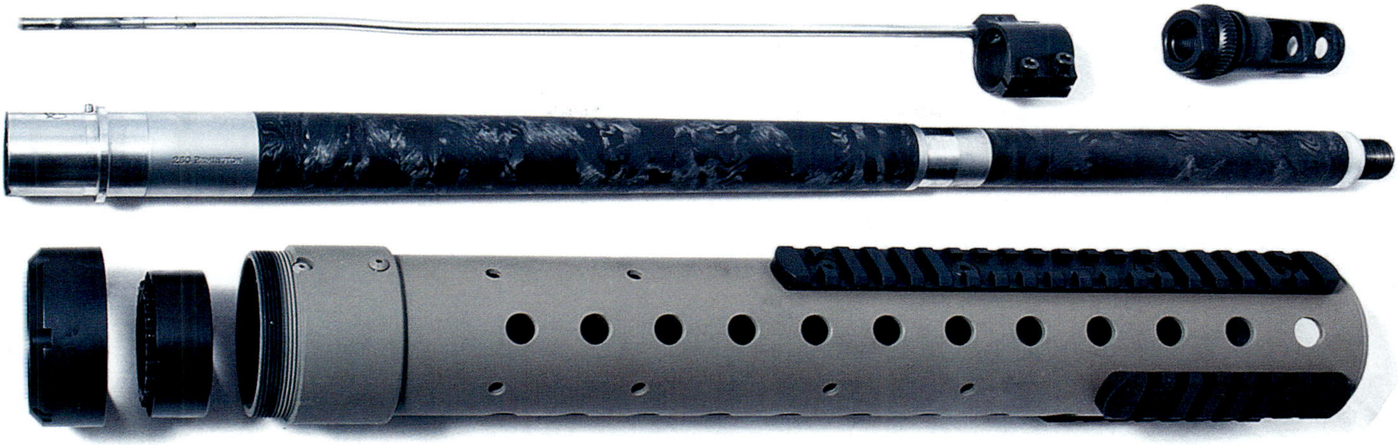

but reach out and touch it.

We'd moved the ball toward our goal of a lightweight rig, but would it shoot? The rifle is capable of sub-½ MOA groups from the bench, but I wanted to see how it would perform under more realistic conditions. I headed to Texas with a Pelican case full of rifle parts, a photographer, and a supply of my editor-in-chief's handloads.

AR-10 disassembled. Despite the barrel's girth, it's much lighter than the original all-steel unit.

REAL-WORLD APPLICATION

Steve Struve of Cawthon Cartridge Club in Millican, Texas, (www.cccshooters.com) gave us the run of his tactical shooting paradise: more than 800 acres of varying terrain and fixed ranges, complete with a 1,000-yard steel range. Range Director Fred Quarles, an accomplished long-range shooter, set us up on one of the training ranges to get our zero. We quickly got the rifle shooting on paper and reset the turrets to zero on the Leupold 4.5-14x50mm ER/T optic before heading to more open country. As we drove to the second range, I reached out and wrapped my hand around the Proof Research barrel that had just sent a couple of dozen rounds downrange both with and without an AAC suppres-

.308 (right) versus .260.
Bigger isn't always better — the smaller diameter, high BC bullet is more effective at long range.

sor — it was cool enough to touch

Fred's ballistic app told us to dial 9 MOA of elevation for the 500-yard target and the first round hit the steel with a satisfying "smack." A little fine-tuning for wind and we were drilling centers with boring regularity. It was time to stretch it out and see if the rifle was up to some truly long-range work.

At 1,000 yards, the 16-inch steel plate taunted me at 14.5x through the swirl of mid-day mirage. This time the app called for 27 MOA up and the 9 o'clock wind had me dialing for double-digit windage.

"Send it."

All eyes were on me as the Armalite's 4-pound trigger started feeling more like 40. The trigger broke and I got the scope back on target in time to see a puff of dirt a few feet to the right of my point of aim. Time to dial more wind.

"Left 7, send it."

Another shot, this time right off target at 6 o'clock, possibly even skimmed the edge of the plate.

"Hold at 1 o'clock, send it."

Exhale, squeeze, bang, recoil... three seconds later, the sound of copper hitting steel.

"Hit!"

I sent another round, again a hit, and so on until we ran out of ammo. We had a legitimate 1,000-yard rifle on our hands, and not some 15-pound custom bolt gun — one that functioned flawlessly from 20-round mags and carried all day easily in an Eberlestock Gunslinger II pack.

LIGHT AND COOL

This rifle is the perfect setup for "run and gun" long-range matches where speed and mobility can be as important as accuracy. And military and law enforcement applications are pretty self-evident. Even though the rifle is light by precision rifle standards, it's still pleasant to shoot, especially with the suppressor attached. You could shoot this rifle all day long without taking a beating.

The gun was light and shot well, but we still had more fat to cut. This rifle would wear a sling and a bipod, but not much else, so I didn't need acres of Picatinny rail weighing me down and tearing up skin and gear. PRI's 15-inch free float tube, made almost entirely from carbon fiber, was a perfect fit for the big Armalite. Weighing only 19 ounces and featuring just enough rail for my intended use, this fore-end was ideal for achieving our goal of cutting as much weight as possible. My SASS was part of a production run that came equipped with FDE furniture (including a UBR stock from Magpul), so the FDE PRI tube kept to the theme. With it installed, our final weight is 12.7 pounds with scope, sling, and bipod. Pretty impressive for a 1,000-yard setup, and less than a pound heavier than the completely nekkid base gun.

When Eugene Stoner engineered the AR-10 using what were space-age materials at the time, he changed the firearm world forever. Decades later, we have access to products and materials that can take his genius to another level without compromising performance. Proof Research has cracked the code of carbon-fiber-wrapped barrels by producing products that cut weight and aid cooling without sacrificing accuracy or durability. The company has a fully funded R&D department that uses hard science to continually innovate its products. Those resources pay dividends when it comes to performance. This rifle is proof positive. R

PROOF RESEARCH BARREL:
Length: 22 inches
Caliber: 6.5mm
Cartridge: .260 Remington
Twist: 1:8 RH
Diameter at gas block: .975 inches
Weight: 1 pound, 13.5 ounces
MSRP: $1,040

FEATURED ACCESSORIES:
Leupold Mark 4 ER/T 4.5-14x50mm (30mm) M3 Front Focal - $2,000
Leupold Integral Mounting Systems Mark 4 - $266
PRI Gen III Armalite 15-inch FDE Carbon Fiber Forearm - $380
AAC Blackout 7.62 51T - $150
AAC Hunter - $650
Harris Bipods HBR 6 to 9 inches - $90
Caldwell Shooting Supplies Picatinny Rail Adapter - $20
Viking Tactics Wide Padded Sling - $41
Magpul UBR Stock - $265

PRICE AS FEATURED: $8,002

ARMALITE SUPER SASS

CALIBER: 7.62x51mm NATO
OVERALL LENGTH: 41 inches
WEIGHT: 11 pounds, 13.44 ounces
MSRP: $3,100
URL: www.proofresearch.com

POUND FOR POUND

LIGHTWEIGHT (CARBINE) AT ANY AND ALL COSTS

STORY AND PHOTOS BY **STEVEN KUO**

Floyd Mayweather Jr. burst onto the national scene as a featherweight, winning a bronze medal in the 1996 Olympics and never looking back. Putting aside all the controversy in his personal life, Mayweather is undefeated in his professional career and his technical brilliance cannot be questioned. He's widely considered to be one of the greatest pound-for-pound fighters of all time.

Bentwood Gunsmithing aims to do the same with their featherweight OIP ("Ounces Is Pounds") carbine — delivering just what you need at the lightest possible weight with the greatest possible performance. Have they succeeded? Let's take a look.

HOW LOW CAN YOU GO

The OIP had its genesis as a simple idea to build a lightweight gun that just plain worked. Dave Lake and Matt Babb of Bentwood Gunsmithing spent years perfecting the concept, incorporating the latest components where they existed and working with companies to customize parts that didn't. They wanted a well-balanced gun with an optimized operating system and literally no excess — to be as light as humanly possible.

Bentwood didn't want to utilize polymer receivers and worked with Battle Arms Development to develop a super lightweight receiver set with intricate machining to shave as much weight as they could. They investigated some more exotic material choices, but found them to be prohibitively expensive. The top rail is abbreviated, making it clear that Bentwood intended this rifle to be run with a red dot optic without backup iron sights — Bentwood's Dave Lake explained that the OIP was designed for domestic professionals and consumers, not military end users. "Better to have them and not need them than need them and not have them" was not a refrain in Bentwood's keep-it-lean design philosophy for the OIP. The upper and lower weigh in at just 12.3 ounces, amazingly light for billet aluminum. You can purchase similar receivers from Battle Arms, though they aren't as aggressively lightened.

The barrel is the heaviest part of a rifle. There are unavoidable functional compromises in designing a barrel, as lighter barrels will give up accuracy more quickly as they heat up than heavier ones — but heavy profiles can be cumbersome boat anchors. Bentwood makes no bones about this and made great efforts to strike a good balance — they were willing to really invest in the barrel. It's manufactured by Faxon, from 4150 steel with a Wylde chamber and nitride treatment. Bentwood went so far as to license Knight's Armament Company's patented dimpling to reduce weight, while maintaining as much stiffness as possible. There's also more metal on the breech end, and the gas journal is sized perfectly for the gas block. Pinned and welded to the barrel is a customized, suppressor-compatible, titanium muzzle brake based on a Hawkins Precision design (a flash hider is optional). The barrel is machined to time the muzzle brake and just long enough to reach 16 inches in overall length. The extra investment in machining maximizes the barrel length and obviates the need for washers and an unnecessarily long muzzle device to avoid SBR status. The barrel alone is about 17 ounces; Bentwood experimented with even lighter profiles, but decided against it.

The handguard is another prime area to cut weight, and Bentwood opted for a 9-inch-long carbon-fiber handguard. It's secured to the barrel nut with an advanced (and expensive) aerospace adhesive, saving additional weight at the expense of not being serviceable — even at an armorer level. Large ventilation cuts adorn the handguard and let you admire the beautiful handiwork below...more on this later.

The bolt carrier group is ion-bonded titanium, made by Toolcraft, and is but one of many other components made of the wonder material boasting the highest strength-to-density ratio of any metallic element. Everything from screws and pins to the impossibly tiny gas block sparkle with that unmistakable titanium sheen. It's not cheap, but it's strong, light, and highly corrosion resistant.

The custom receiver extension and skeletonized butt plate round out the ass-end of the rifle, housing a unique buffer system with a dual rate Hilsabeck Firearms action spring and a custom hollow leaded steel buffer. In combination with the gas system and titanium bolt carrier group, the system is tuned to minimize reciprocating weight while mitigating the recoil impulse from the buffer hitting the end of the receiver extension and the secondary impulse of the bolt going back into battery.

The titanium gas block is absurdly tiny, and the Battle Arms Development upper and lower are intricately machined to remove a lot of weight. Anything that can be skeletonized or made of titanium has been made so.

The fire control group is the excellent JP Enterprises single-stage (which broke cleanly just under 3.25 pounds), and the safety selector is a slick-as-snot skeletonized Battle Arms Development unit. An optional Picatinny light mount attaches to the Fortis Aimpoint mount and only weighs 1.1 ounces.

INSERT HAN SOLO/TOY GUN JOKE HERE

You have to handle the OIP yourself to truly appreciate just how crazy light it is. Without an optic, it weighs 3 pounds, 14.5 ounces. With the Aimpoint H-1, it's just 4 pounds, 3 ounces. Its unconventional looks combined with its inconceivably light weight honestly makes it feel like a toy when you first pick it up. But once you send some rounds downrange, you'll be checking if you're on Santa's nice list.

One thing is clear, the OIP doesn't really want to be shot from a bench. The receiver extension's too short, the buttstock doesn't help at all, the top rail is tiny — nothing about the OIP is conducive to it. But we've gotta start somewhere, and that's usually on the bench, to zero, group, and chrono the gun.

Bobro to the rescue, as their clever and compact mounting system was able to clamp onto the abbreviated top rail and allowed us to mount a traditional magnified optic for bench work. Not that we'd really want to run this kind of glass on the OIP, but it's helpful for our testing.

Surplus Ammo supplied us with their 55-grain FMJ new production ammo, which averaged 3,016 fps out of the OIP (at 87 degrees and 2,400 feet) and was our primary test ammo for the OIP. We don't sugar coat things here at RECOIL, and we also don't pretend to be the anointed hand of Thor, raining sub-MOA hate down upon any and all targets within 1,000 yards with any weapon system that crosses our path. Sure enough, off the bench, we saw 2 to 3 MOA groups with the 55s. So we figured we'd try a few groups with some match ammo — our best groups were 1.7 MOA with 75-grain Black Hills and 1.4 MOA with 69-grain Gorilla SMKs, but some opened up further.

We reinstalled and zeroed the Aimpoint H-1 and couldn't get away from the bench fast enough to shoot a match with the OIP.

HOSEFEST-PALOOZA

Once we started running and gunning, the OIP really came into its own. The rifle is so light that it's no surprise that transitions are lightning quick and you can drive the gun easily. What is a surprise is how mild the recoil impulse is and how flat the gun shoots for such a featherweight.

Bentwood has really done a great job tuning the gun (note that they don't guarantee reliability with all types of ammo). We've spent many a long range day tuning the AR platform — adjusting gas systems, trying different types of bolt carriers, swapping out various types of buffer systems, experimenting with different muzzle devices, and quite often cursing and bitching as a result. So we know firsthand the intricacies of tuning the operating system and how tricky it can be, especially when running at the edge of the performance envelope.

With the OIP's light weight and soft recoil, we successfully engaged steel at 100 yards holding the rifle

one-handed like Han Solo's blaster. It was pretty straightforward once you got the hang of it. All fun and games until you realize that every firearms trainer teaches one-handed weapons manipulation — for good reason, because bad things happen in gunfights. And anyone who's done those training evolutions knows how challenging they can be with a typical rifle if you don't look like Buck Doyle (see page 166). It's night and day with the 4-pound OIP. Yet, the rifle was still easily controllable when hammering targets and dumping mags. Mount it securely — one of our test shooters was sloppy about this and complained about the bare metal buttstock penalizing him for it. Taller shooters were very cramped behind the gun, but it's tons of fun.

We loaded up a bunch of magazines with the Surplus Ammo 55-grain offering and took the OIP out to an action rifle match, featuring a mix of fast and furious hoser stages and long-range stages with steel targets out to 360 yards. The little blaster was a delight on the short courses, swinging to targets almost telepathically and double-tapping them easily with a flash sight picture — it even notched a stage win in Limited division on one of them. It acquitted itself honorably on the longer stages, easily ringing mini poppers out to 175 yards, and hitting 10-inch flashers from 200 to 315 yards. The final 360-yard flasher went untouched, as this scribe's aging eyes just couldn't sufficiently zero in on the target, less than 3 MOA in size at that distance. That frustration aside, we were pleased to get our hits with the tiny OIP, red dot, and non-match Surplus Ammo 55-grain loads. The rifle was a real blast and performed well. We passed it around the squad, and even the goliaths enjoyed shooting the tiny little gun.

But nothing is without some gripes, so here are ours. The OIP's custom buttstock is roughly the length of a typical fully collapsed stock. Bentwood asserts it's intended to be used in (and encourages) a fully squared-off stance. If you fall in that category, perhaps it'll work for you. But every one of our shooters felt it was too short; we'd happily take the weight penalty for a longer buttstock. The 9-inch handguard is perfectly serviceable, though many of us would have preferred a longer one. But chalk that to preference — what was unequivocally a bad idea was the huge ventilation holes in the handguard right above the flaming hot gas tube. You know where this is going; two shooters burnt their fingers on the gas tube — one when handling the weapon, so he escaped an annoying injury, and the other during a course of fire, so he wasn't so lucky. Bentwood has since redesigned the handguard to delete the aft-most hole, but retained the forward one. Unfortunately, the forward hole was the one responsible for the burnt finger. Bentwood is already looking at other options for future revisions. Until then, keep your gloves on.

The light mount is handy and barely over an ounce, but it attaches to the scope mount, putting it high on the rifle and close to the receiver. This results in nice overall weight balance, especially given how light the rifle itself is, but it essentially necessitates a magwell grip to activate the light and can interfere with grasping the handguard forward of the light. The X400 on our test gun was convenient to use with a magwell grip, so those who use that technique will like it — but most of us here aren't among them. It's also not an ambidextrous setup, and the handguard doesn't otherwise accommodate attachments.

The OIP has clearly been very thoughtfully and carefully designed as a complete system, and it works well. To achieve the immense weight savings, Bentwood had to stick closely to a specific design philosophy. Mayweather's not for everyone, but he's a brilliant boxer and unquestionably entertaining. The OIP is similar — if you agree with all of Bentwood's design choices, you'll love it. If you don't, you'll definitely wish they did something different here or there, but you can't help but appreciate the overall package.

And we bet you'll have a big smile on your face after sending rounds downrange with one. RV

The optional 1.1-ounce picatinny light mount attaches to the customized Fortis Aimpoint mount. It places the light (Surefire X400 shown here) rather high and close to the receiver. If you like a magwell grip, you'll like its placement.

The buffer system utilizes a dual rate Hilsabeck Firearms action spring and a custom hollow stainless steel buffer. The customized Toolcraft BCG is titanium. Even the magazine catch has been shaved down to the minimum needed.

BENTWOOD GUNSMITHING OIP CARBINE

CALIBER: .223

MAGAZINE CAPACITY: 30 rounds

BARREL LENGTH: 16 inches (permanently attached muzzle device)

OVERALL LENGTH: 32 inches

WEIGHT: 3 pounds, 14.5 ounces

MSRP: $2,879

URL: www.proofresearch.com

FEATURED ACCESSORIES:
Accessory kit (Aimpoint H-1, Fortis mount, three extra magazines, single point sling, carry case) - $775
Light mount - $TBD
Surefire X400 Ultra (Green Laser) - $790

PRICE AS FEATURED: $4,444

FULL-SCALE ASSAULT

Erathr3's Project Anorexia Puts the AR on a Diet

By Rob Curtis

All weight-loss programs aren't the same. While both may weigh 110 pounds, there's a difference between a methadone addict and an Olympic athlete. No more would we trust a meth addict to function in a life-threatening situation than we'd expect a Swiss cheese AR to stand up to the rigors of combat.

And, meth addicts always become ugly. Nobody wants to date a meth addict.

WAIT, WEIGHT

We spent three days running a pre-production Project Anorexia rifle at Thunder Ranch with one of the driving forces behind the gun's inception, Clint Smith. Our test rifle was outfitted with a scope and bipod according to Smith's specifications for use in his urban precision rifle class. The class turned out to be the ideal venue to highlight the advantages the skinny gun he helped design brings to the intermediate distance fight.

BEN DITTO PHOTO

"I designed this gun mostly for women and myself, because I don't want to carry heavy shit any more," Smith said about the gun, "If I'm going to carry 2 extra pounds of something, it isn't going to be a barrel. It's going to be 2 pounds of ammo."

Smith's longtime friendship with Erathr3's Sheri Johnson led to the collaboration. It was a great match for this project, with Erathr3's passion for high performance tempered by Smith's experience. The combination meant the rifle was built with no quarter given to price point, only performance and reliability.

Getting a 6.5-pound flattop AR is hardly a challenge these days, but building an accurate, reliable gun that weighs less than 6 pounds will take some talent.

ANOTHER AR

"We didn't enter thinking we were doing anything earth shattering," said Erathr3's Sterling Becklin. "We dig them, we use them all the time, so we wanted to make our own reliable, accurate rifle. It's just something we wanted to do."

Becklin says Erathr3 started with the barrel and built the gun around it. "I believe in Proof Research's technology," explained Backlin, "not just the carbon fiber, but the way they're CNC rifling and chambering."

The 1/7-inch twist, 14.5-inch Proof Research carbon-fiber composite barrel headlines, but the ensemble cast of supporting actors steals this lightweight show. In order to shatter the 6-pound AR barrier, Erathr3 had to get into the weeds. All the small steel parts were replaced with lighter weight titanium parts. The only exception is the traditionally steel ejection port cover. Instead of titanium, it's aluminum — lighter weight and perfectly acceptable for this unstressed part. But, the ejection port cover hinge is titanium. The only small part left in a legacy material we could find is the bolt catch; and even that is V Seven's lightened version made from super-duty S7 tool steel.

While some larger components got the Ti treatment, some had to remain as good old ferrous alloys. But that doesn't mean they were left in the 1960s. JP Enterprises steel JP EnhancedBolt adds material around the cam pin hole and uses JP's enhanced ejector, extractor, and gas ring designs. JP also uses SAE 9310 steel in place of Mil-spec Carpenter 158 steel; it's something the company says improves the life span of the bolt. Erathr3's proprietary steel barrel nut uses integral wrench flats that also shave a little weight.

Bolt carrier, takedown pins, trigger pins, buffer retaining pin, fire control selector, grip screw … all titanium. But one set of Ti parts offered a unique challenge. "Yeah, the first time we tried staking the titanium castle nut, the titanium end plate just chipped," Becklin laughed, "so we tried different tools and dicked up two more before we came up with a solution." They used an aviation industry staple, Torque Seal.

GIVE IT THE FINGER

Erathr3's receivers are made in house, allowing them to fine tune everything, including the skeletonizing pattern and the lightening cuts. Running the "build, assess, improve" cycle in real time results in machining tolerances so fine they even account for the thickness of the gun's gray Cerakote finish and result in parts that fit together a like a pair of scissoring college sophomores.

The lower receiver's forced weight-loss program features skeletonizing of the flared magazine well that does a couple things. Not only does it reduce weight, it also lets a shooter grab or sweep out hung-up casings with a couple of fingers; something that should be done with care, given the amount of force that will be applied to flesh if the bolt is allowed to slam home.

We ran the gun with a Spuhr Mounts special edition Erathr3 3026 mount that's lightened and has a titanium clamp. It's shown holding the Schmidt & Bender 1.1-4x24mm Zenith Short Dot LE we used.

The titanium endplate and castle nut can't be staked, so Erathr3 used a dab of orange Torque Seal as a witness mark.

The trigger guard was scalloped a bit aggressively. It was a sore subject and one that Erathr3 has already corrected.

More than for just looks and lightness, the magwell cutouts provide finger access to stuck casings.

The upper also goes on a gag-reflex diet, losing the forward assist and material on the left side with some strategic scalloping. A few more ounces come off with a Ti gas block and Proof Research's CF composite barrel. The weight savings between a 14.5-inch steel barrel and a CF composite is minimal; the weight delta is much more obvious when comparing 18-inch steel and CF barrels. But, E3 wasn't trying to build on a budget. CF is stiffer than steel, which can translate into a more accurate barrel (if you buy into that whole science thing), so adding performance while removing even a small amount of weight fits the Project Anorexia MO.

Just because the gun is made entirely from high-quality parts doesn't mean the assembly process is a carefree wrench-athon. Parts are still carefully fitted. For instance, before declaring a barrel and bolt soulmates and locking them together in metallic matrimony, Erathr3 gauges the fit and selects the bolt with the best headspace for that barrel and extension assembly.

The basic Anorexia build is rounded out with the thin-diameter, M-Lok-compatible E3 Handguard; Magpul's stock, grip, sling; MBUS Pro offset iron sights; AXTS Raptor charging handle; Inconel gas tube; a pinned SilencerCo Trifecta flash hider; and an ALG Defense ACT trigger. We'll talk about the trigger in a minute...

THE THUNDER RANCH EXPERIENCE

The limited edition of the carbine we shot was outfitted specially by and for Clint Smith's urban rifle course that served as our three-day, 2,000-round introduction to project Anorexia.

Smith wanted to show the lightweight, 14.5-inch, 5.56mm platform was an effective tool for precision work in an urban setting where targets are seldom more than 300 yards out. To that end, our gun was set up with a Schmidt & Bender 1.1-4x24 Zenith Short Dot LE and a BT10-LW17 V8 Atlas Bipod.

We went from zeroing optics on day one to ringing small steel as far as 400 yards from rock piles, rooftops, and truck beds at the end of day three. In the interim, we worked on precision, positional shooting, movement, shooting movers, and working 100+ yard targets from the awkward, external apertures in Smith's signature *Terminator* shoothouse. Many of the drills involved stressors that helped point out how much a lightweight gun helped manage fatigue. Despite Anorexia's svelteness, "Hold that gun up," was a still painful Smith refrain.

While the schoolhouse experience and subsequent testing showed the gun was reliable and accurate (our best five-shot group was 0.57 MOA using Black Hills blue box 77-grain MatchKing HPBT), we're pretty sure we couldn't access the gun's benchrest accuracy potential owing to the ACT trigger. Smith spec'd that trigger himself for the gun. He told us he wanted a no BS 4-pound combat trigger with solid reset and he got one. The ACT is built to improve on the 1- to 2-MOA performance of rack grade rifles, not wring the last .25 MOA out of a $3,700 custom build. That said, Anorexia isn't a bench gun, and we could reach past 300 yards without trouble. In the end, Erathr3 understands the market's desires and will offer Anorexia with an option for a higher-grade bang switch.

GRIPES

It's a $3,700 gun. There shouldn't be any gripes. But, aside from the trigger, we noted two issues. In the quest for the lightest lower, Erathr3 scalloped the bottom of the trigger guard all the way to the grip. The ridges, even rounded as they are, chafe the hell out of the shooters middle finger after a few magazines.

The other problem would barely be worth mentioning if this weren't a premium rifle. The handguard has what looks like M-LOK cutouts on the four cardinal faces. Turns out, only the three, six, and nine o'clock cutouts are for M-LOK. The top readily accepts M-LOK parts, but there isn't enough space on the backside to turn an M-LOK cam into its torqued position. So, guys that try to mount something up there, say an offset light or action cam, are going to think it's securely mounted right up until the device falls off, as ours nearly did.

The company is aware of both issues and has already recontoured the trigger guard for the next run of lowers and changed the rail cutouts to make it more obvious that the top is not an M-LOK mount point. This points back to one of Erathr3's strengths, it can make changes like this on the fly because the company machines its own parts.

BEN DITTO PHOTO

ZIPPING IT UP

Packing the rifle in its accompanying Grey Ghost Gear rifle case for its shipment from our offices, we've felt something rare when saying farewell to an AR. Usually, we're thankful to get the space back in the safe when we're done... but, we're going to miss this one when the brown truck pulls away. Few ARs come as close to the perfect balance of performance and weight as Anorexia. It was a fun gun to shoot and never once did we get the feeling that there was some hinky, lightweight part used to shave an ounce.

We can't be too sad, though. This one, Anorexia 16, is going to AmericanSnipers.org to be raffled off as a fundraiser that will begin at SHOT Show 2016. $10 will buy you a chance to win this gun, complete with optic, bipod, case, and a few PMAGs. And, your $10 goes to buy equipment that downrange snipers need ,but can't get through the normal supply chain. More info will be available at AmericanSnipers.org. R

ERATHR3 PROJECT ANOREXIA

LIMITED-EDITION CLINT SMITH SIGNATURE "SKINNY GUN"

CALIBER: 5.56 NATO

MAGAZINE CAPACITY: 30 rounds

BARREL LENGTH: 14.5 inches

OVERALL LENGTH: 31.75 inches collapsed (35 inches, open)

WEIGHT: 5.7 pounds, rifle only. 8.4 pounds as shown

MSRP: $3,750

URL: www.erathr3.com

FEATURED ACCESSORIES:
Schmidt & Bender 1.1-4x24 Zenith
Short Dot LE (w/CQB Reticle) - $2,754
Atlas Bipod - $280
EraThr3 Skeletonized SP-3026 Mount by SPUHR - $475

PRICE AS FEATURED: $7,250

LIGHTWEIGHT, NERDY, AND DIRTY

Diving Deep into the Enyo from Master of Arms

By David Merrill

In the time since Eugene Stoner first started developing the Armalite Rifle back when Chubby Checker was at the top of the charts, a whole lot has happened to the system. Not even Stoner knew that it would truly become the modular rifle of the future, rather than his later pet project specifically intended for the purpose, the Stoner 63.

Barrels got shorter, gas systems moved, silencers layered in complications, calibers changed, controls updated and swapped, belt-fed uppers constructed, heavier and lighter bolt carriers shaped, specialty bolts manufactured, extractors revised — there isn't a single piece of the AR, from the smallest spring to the longest barrel, that hasn't been changed, modified, and otherwise refined over time.

As a gun owner in 2017 it's not hard to take a whole helluva lot for granted — every inch of progress gained started out as something design firms like to refer to as "challenges" rather than the more direct and honest word "problems."

However, not every change could objectively be called an improvement. Many initially promising alterations later became relegated to mere oddities and gimmicks, or worse, hindrances. As the saying goes, you have to break a few eggs to find a diamond (or something like that).

In recent years, lightweight has been the name of the game. Blindly running down that rabbit hole can lead to absurdity and the indiscreet burning of copious amounts of cash. Zero in on a single goal over all others, and you'll end up with some insane monster. #SkeletonizeEverything

Even so, the most satisfying endeavors are never the easy ones. And looking over the Enyo from Master of Arms gives an inkling that this rifle was a labor of love, obsession, absolute nerdery, and yes, a dab of insanity. What we have in our hands is a preproduction prototype, so there are bound to be at least some minor changes when it's in full production. We'll note these as we go through the rifle.

We'll state outright that this is the about farthest thing from an "Everyman" AR as one can get. The Enyo is costly, nearly entirely proprietary, requires special tools to work on, and an intimate and nuanced knowledge of the AR operating system is advisable.

NOTHING NEW, AT LEAST NOT ON THE SURFACE

Initially, it looks fairly straightforward. From the minimalist stock and all of the carbon fiber, it's plainly obvious the Enyo had light weight in mind. There's also the mandatory skinny barrel. But we've seen lightweight before, several of them in fact. Back in Issue 21, we covered the 3-pound, 14.5-ounce (naked) Bentwood Gunsmithing OIP carbine. Further in that same issue we even built up our own from spare parts, coming in right at 4 pounds without sights or optics. At the 2017 SHOT Show, we played with a 3.5-pound Raptor AR from Fostech that we could easily swing from our pinky. So while somewhat rare in a package that weighs under 4 pounds, a lightweight AR in and of itself is nothing new.

DIGGING IN

The upper and lower receivers on the Enyo aren't your standard fare. Sure, they have the sleek weight-saving designs you see with many a billet receiver set — gone is the cage around the magazine release, port door, forward assist, and brass deflector. Even the top rail has a radius taken out of it. There are other weight saving measures too, such as the lack of a traditional castle nut/endplate system and even a (previously) ambidextrous safety selector with one side taken off.

It's obvious as soon as you pick it up that you're not dealing with your typical 7075-T6 aluminum, even a severely skeletonized one. And while Master of Arms does indeed sell a similar set to this one in 7075,

Exotic materials and lightening cuts make for a distinct look.

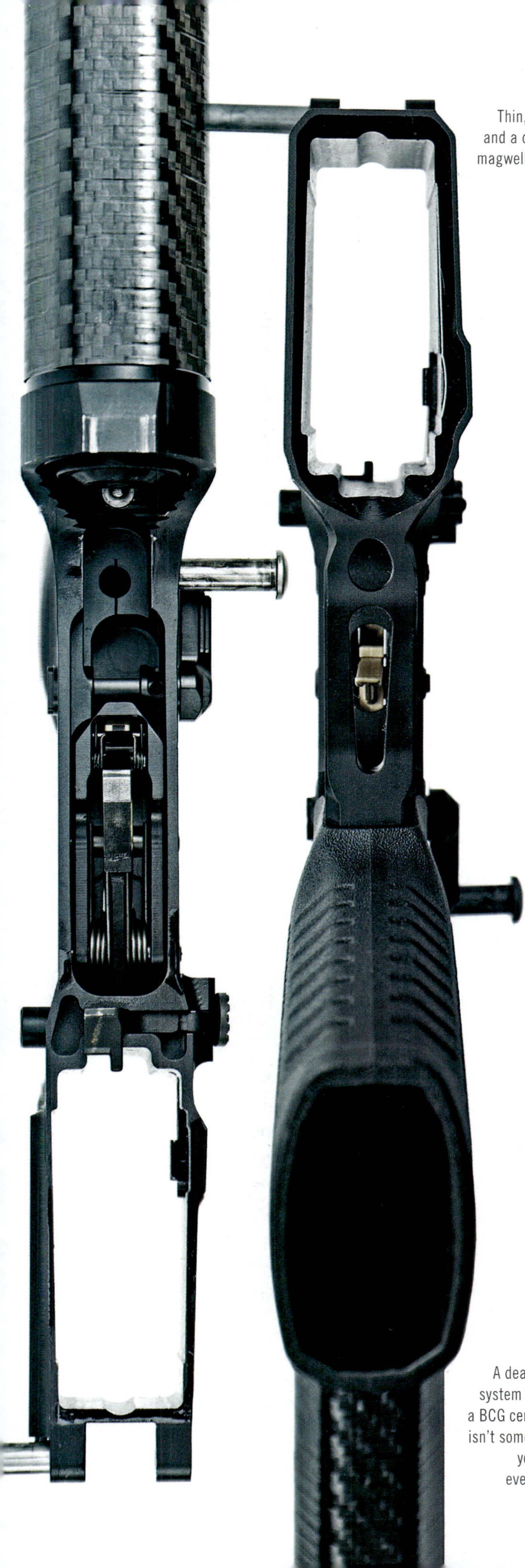

Thin, light, and a decent magwell flare.

dubbed the Esoteric, the Arcane set on the Enyo is something entirely different.

In the past we've seen plastic receivers, carbon-fiber forms, and various magnesium alloy sets with various degrees of success. Master of Arms chose something different: 7095 Tennalum.

This alloy has a strength-to-weight ratio approximately 35-percent higher than 7075-T6 and 28-percent higher than Grade 5 titanium. Of course, this means that it's expensive and thus rarely seen in our market. We do question the wear resistance of 7095 Tennalum, as there's a steel pin insert to secure the charging handle latch. However, this should be noted purely as speculation at this point.

Crack the rifle open, and it gets even weirder. Instead of pushing both takedown pins out left-to-right, the rear removes in the opposite direction. When we asked Master of Arms about this, we expected some kind of method to their madness. Instead, we got this: California laws, man, California laws. You see, under the current terrible restrictions in the Golden State, reloads with an AR may require the receiver be partially broken down to accomplish, depending on how you have the rifle configured. Having one pin go a different direction makes this slightly easier for their California customers who are lefties.

Thankfully, our Enyo required no such silly process to reload. F*ck California gun laws. There's also a nice flare built into the magwell to ease reloads.

The inside view of the Arcane receiver shows some expected weight saving measures as well; nothing drastic, just thinner walls than normal and pockets milled out where a traditional receiver would sport solid metal.

A deadblow system inside a BCG certainly isn't something you see everyday.

BOLT CARRIER GROUP

There are two newly engineered and somewhat bizarre configurations in the Enyo that initially caught our attention, and if we're being honest, our imagination. The first is the Nyx Titanium carrier. The engineers at Master of Arms began with the standard dimensions of a Mil-spec BCG, and then cut as much weight off as possible. They then machined it out of Grade 5 titanium, and most curiously, added a secondary dead-blow system to the carrier itself.

The Enyo shipped with balls of various weights and materials, accompanying springs, and the tools to replace them, all for use with the secondary dead-blow system.

We already know your next question: So what the hell is this supposed to accomplish?

One universal issue we've seen with featherweight AR-15s is that while they're easy to carry, they become a bear to shoot. Even the lightweight 5.56 round can beat your ass if you're not careful with your recoil management; there's just not enough mass in the rifle to absorb perceived recoil impulse. This usually leads to loud-assed muzzle brakes. Master of Arms didn't want a brake (more on the why later in the article) so the secondary dead-blow system was developed.

Though the rifle will run with the Nyx BCG empty, assuming you have your gas tuned properly, the addition of the tungsten, brass, or aluminum balls allows the recoil to better move directly into your shoulder instead of up. This equates to less dot or reticle movement from shot-to-shot. At the time of this writing, while the titanium ball works the best, it also wears out springs much faster. Master of Arms is currently experimenting with different spring configurations to increase the lifespan when used with the heaviest ball.

We can confirm that without the aid of the secondary dead-blow system, perceived recoil is very sharp. With the tungsten ball in place, it's still worse than a standard M4, but

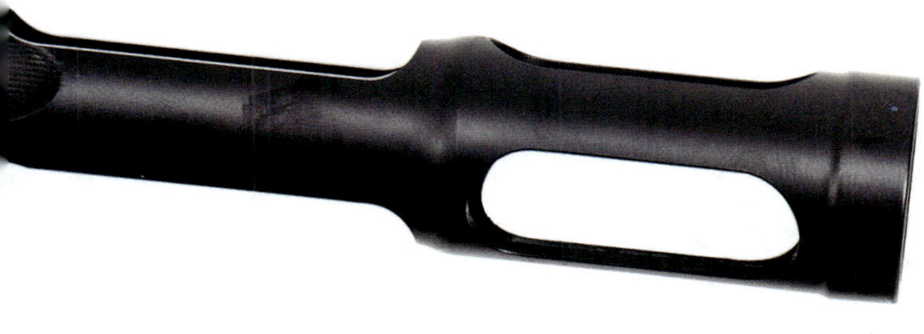

Some tools required. Dumbass dirt shooters need not apply.

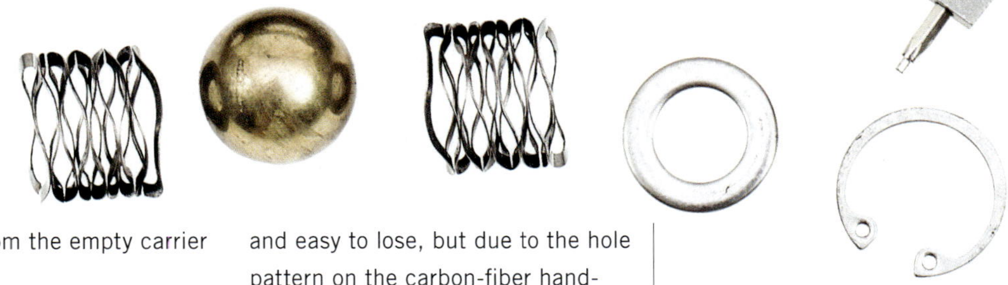

is far blunted from the empty carrier arrangement.

Replacement or removal of the ball and springs requires special tools. Have we said this wasn't a rifle for beginners?

Of course, one can also play with the buffer and spring arrangement. The Enyo ships with a Cobalt Kinetics buffer system so you can get as weird as you want.

GAS BLOCK

The Enyo ships standard with an adjustable titanium gas block. A single screw with a Bellville washer clamps the block to the barrel. In theory the gas block is infinitely adjustable, sporting two screws: one to make adjustments, and another to lock it in place. Designs like this usually require a replacement of the screws at some point due to the hot gas from the barrel eroding them over time.

Have you ever done something you've immediately regretted? Maybe it was just one too many tacos at dollar taco night, or perhaps it was engaging in unprotected intercourse with a sex worker in Southeast Asia. Unintentionally removing the locking screw of the adjustable gas block falls somewhere in between those two situations. Not only is it small and easy to lose, but due to the hole pattern on the carbon-fiber handguard itself, it was virtually impossible to replace without removal of the handguard. Which brings us to our next topic ...

BARREL AND HANDGUARD

The barrel and the handguard of the Master of Arms Enyo should be viewed as a married pair. With ordinary AR-15s, replacing a handguard is fairly routine. This isn't the case with the Enyo. Master of Arms started out with the skinniest pencil barrel they could find commercially (16-inch Faxon Pencil Barrel) then inexplicably made it thinner. Is ultra-pencil a thing? It is now. Master of Arms tells us commercial models will feature in-house made barrels.

There are a couple problems with pencil barrels: they're less stiff so they whip and flex more than a thicker-profiled barrel, and their thermal mass is reduced so shot groups open up faster.

As with other downsides to the ultralight AR, Master of Arms tried to tackle this issue in an unconventional way. There's a bigassed Torx nut at the end of the handguard. While initially we thought this might be another Californianism and just a creative way to not have a threaded barrel, it's something else entirely.

The entire barrel assembly is held in tension with the handguard in order to increase stiffness without

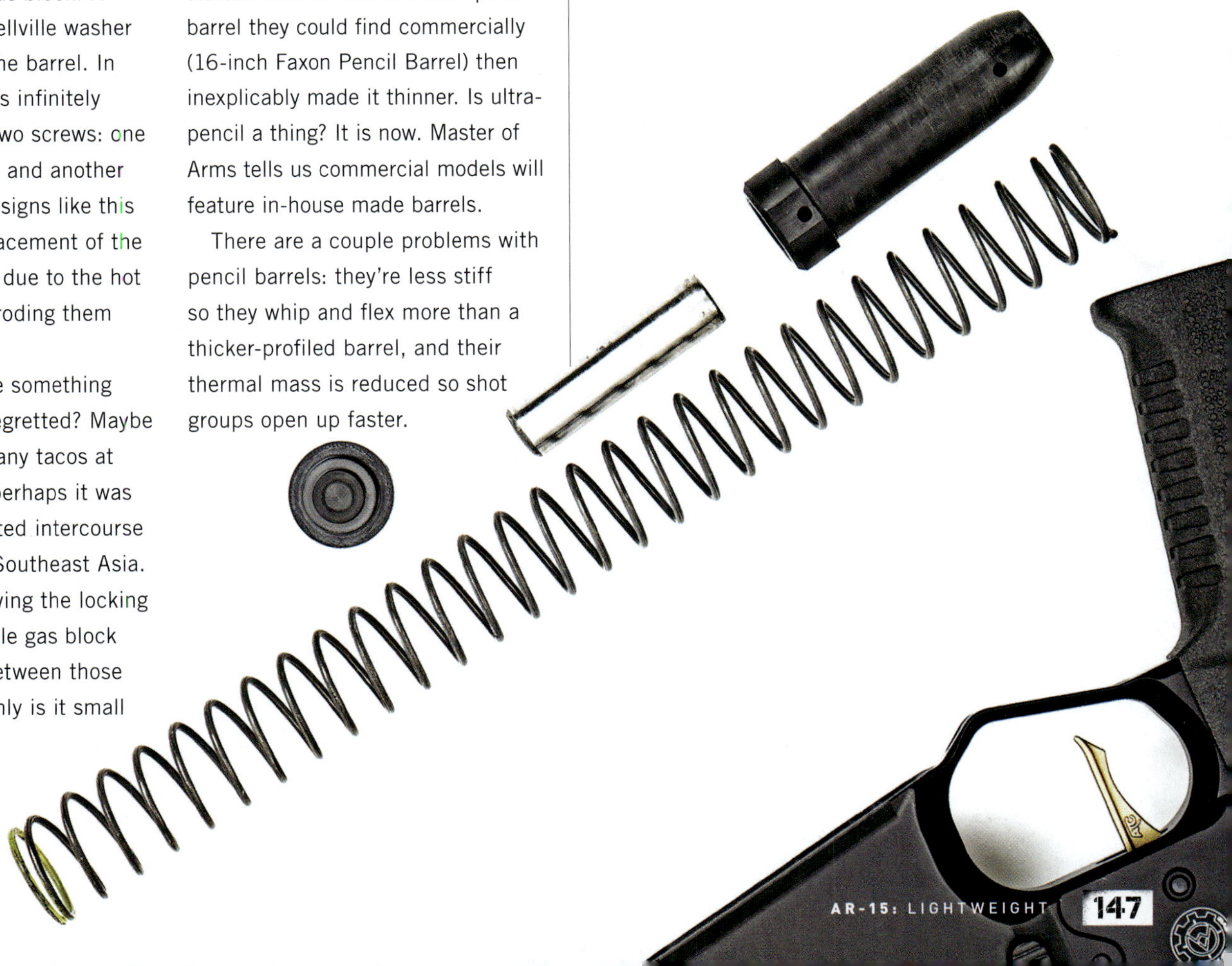

thickening the profile. Master of Arms heats up the barrel to ensure maximum thermal expansion, then torques the handguard cap onto the threaded barrel. Once cooled, the barrel contracts and adds tension. This also means they don't require the use of thread locker to keep the handguard cap on.

Exactly how much tension can be added with a carbon-fiber handguard over an aluminum or magnesium alloy model is unknown at this time, though Master of Arms went out of their way to tell us production models will utilize a different manufacturer of carbon fiber; if you look closely you can see some warps and swells from strain.

In order for a proper tensioned versus no tension test to be performed, it would require the handguard be removed. We were told in no uncertain terms that this shouldn't be done by the end user. No switching from M-Lok to KeyMod to Picatinny to PCAP here.

SO HOW'S IT SHOOT?

Originally advertised with a skeletonized Elftmann trigger, the example we have sports an AR Gold with a pull weight of under 2 pounds from the bottom of the straight trigger. A 45-degree safety throw is also a plus. Magazines locked in the way they should, and rounds fed smoothly from the magazine to the point it sometimes felt like a dry charge. There are some concerns about not having a fence around the magazine release (it was added for a reason), but we didn't experience any premature releases. Perhaps a carbine class or something more vigorous than a testing session might rear its ugly head.

As expected and mentioned above, the recoil impulse is sharp but can be lessened somewhat by utilizing the secondary dead-blow system. Experimenting with ammunition selection, gas volume, dead-blow ball weights, buffers, and action springs eventually leads to the most comfortable shooting ultralight on the market. Granted, your envelope of ammunition is severely restricted here, but this ain't a rifle to bring to war.

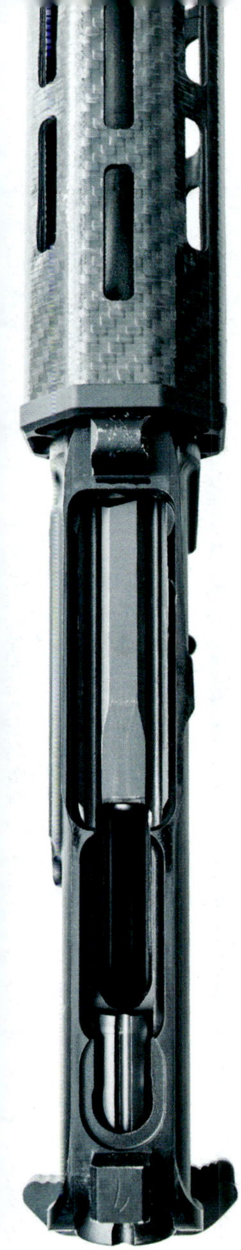

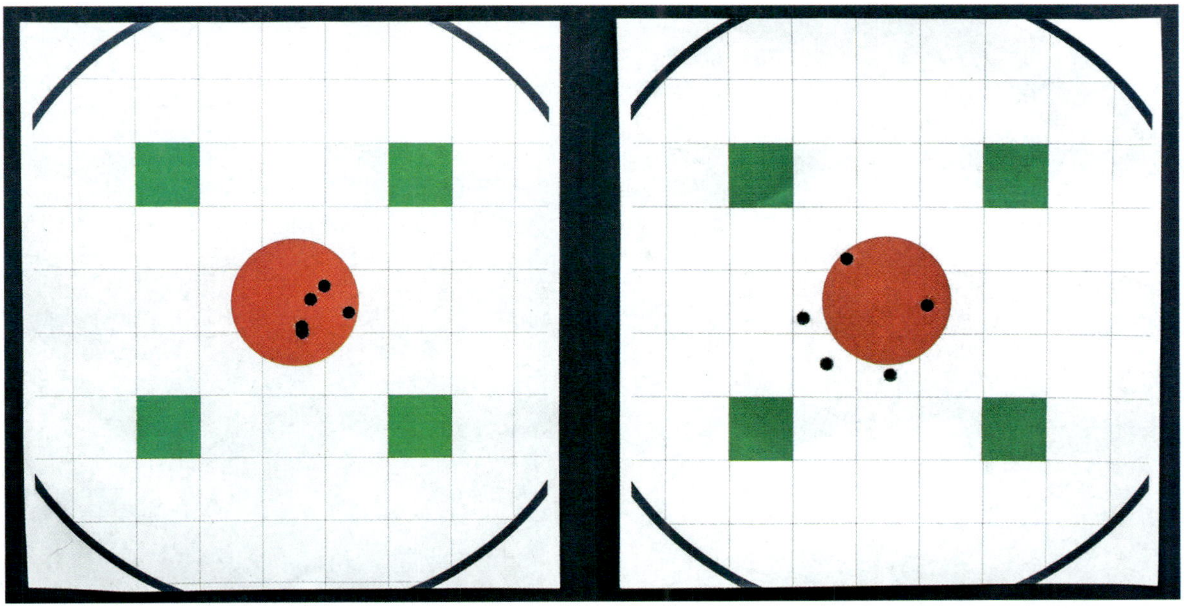

For accuracy testing we took off the micro red-dot and mounted something a little more appropriate to the task: a SIG SAUER Tango4 4-16x optic. After a rudimentary zero we put it on paper at 100 yards. Using 75-grain Prvi Partizan BTHP match ammo, the Enyo was able to group admirably just below 1 MOA — at first. Within the span of a 30-round magazine, the heat mirage from the ultra-pencil barrel began obscuring the target and groups opened up tremendously.

LOOSE ROUNDS

The Master of Arms Enyo is the nerd's nerd rifle. If you like to test and tinker and lose sleep over the ultimate weight savings and performance equation — this one's for you. If you want to just go to the range and shoot, something more pedestrian may be in order. The gas block left us fairly dissatisfied; we'd prefer something a little easier to configure.

With that said, the Enyo can safely be looked at like a concept car, or even a model strolling down the runway. Though not everything you see will be the most practical when together as one, little bits and pieces may well find a home with a working gun or two. There's great potential for the Nyx BCG in slowing carrier speed of an overgassed short-barreled rifle. The handguard arrangement may see specific use with pencil barrels in the future (*cough* please make one with a threaded barrel *cough*). And the currently exotic-and-expensive 7095 Tennalum may turn out to be *the* lightweight receiver material down the line. R

MASTER OF ARMS ENYO

CALIBER: 5.56 NATO

MAGAZINE CAPACITY: 30 rounds

BARREL LENGTH: 16 inches

OVERALL LENGTH: 33.5 inches

WEIGHT: 3 pounds, 11 ounces

MSRP: $$3,700

URL: masterofarms.com

INCLUDED ACCESSORIES:
Vortex Razor Micro Red-Dot Sight
Battle Arms Development Skeletonized Mount

6-LB HAMMER

Sionics Weapon Systems' Lightweight Patrol Rifle

By Tom Marshall
Photography by Kenda Lenseigne

Obesity is a major health problem in America. We all want to lose weight, right? The AR-15, a rifle that has undoubtedly taken on a personality of its own, is no exception. In fact, in recent years "the black rifle" has had much more success slimming down than this author has.

Do you remember what the average AR looked like just last decade? Miles of quad-rail, lasers the size of TV remotes, D-cell Maglites — you get the idea. In the last couple of years, there's been a distinct shift back to the KISS concept (keep it simple, stupid), of which we're big fans. There's also been a resurgence of interest in AR ultra-lights. This is evidenced in things like carbon-fiber-wrapped barrels, modular direct attachment systems, such as KeyMod and M-LOK, and the use of polymer in everything from flip-sights to lower receivers. Equipment selection is a form of mission analysis. Gear up based on needs, not wants. Efficiency is king and less is more — ammo and alcohol notwithstanding.

Sionics Weapons Systems exemplifies this principle, not just with its latest rifle, but with how it runs its business. This small, agile operation runs heavily on customer feedback. We walked into the shop a few weeks ago with a list of specs for a rifle build — a sub-6-pound rifle. The Sionics folks said, "We're not 100-percent sure we can get below 6 pounds, but we can get close." Then they asked us to wait a bit, because they were working on something that might fit the bill. Less than a month later, they released the Patrol LW, a 16-inch mid-length gas rifle weighing in at 5 pounds, 14 ounces. For reference, my son was born at 8 pounds, 11 ounces. Given the choice, perhaps my wife would rather have carried this rifle around for nine months. After all, it weighs less and doesn't cause morning sickness.

LIGHT EQUALS MIGHT

So, who benefits from the lightweight AR concept? Just about anybody. If you hunt with an AR, it's easier to lug long distances over terrain. If you shoot three-gun, a light rifle will drive faster on multi-target hoser stages. If you're preparing for the worst, it means having one rifle that everyone in the family can shoot. If you work the street or the 'stan and you have to hang a bunch of tech off your handguard, it can help manage

the burden. But make no mistake, the Sionics Patrol LW is not light on features. Off the rack, it sports a B5 SOPMOD stock, Ergo pistol grip, VLTOR Gunfighter charging handle, and a Midwest Industries 15-inch SSK forearm. The Midwest handguard is especially nice. It's KeyMod compatible, with an outside diameter of only 1.5 inches. The fact that it runs a full 15 inches accommodates all shooting styles and protects the pencil-profile barrel from any potential damage or flexing when shooting off benches or barricades.

The heart of this rifle is Sionics' new lightweight barrel. It's not fluted, but has a very thin profile. Its diameter is 0.637 inch behind the gas block port and 0.573 inch in front, and it uses a 0.625-inch block. The CMV4150 steel is Melonite finished both inside and out. Every Sionics barrel (yes, every one) is air gauged, magnetic particle inspected, and X-rayed before leaving the factory. It's chambered in 5.56mm, complete with M4 feed ramp extension and a 1:8 twist. The test gun was capped with a standard A2 birdcage, but other muzzle devices are available on request. For the do-it-yourselfers, this barrel is available standalone from the Sionics website, weighing in at 1.375 pounds.

Behind the barrel sits a bolt carrier group that is MPI and pressure tested, as well as coated with a nickel-Teflon finish. The trigger is an ALG QMS, which is not one of this author's favorite units (e.g. the 4.5-pound CMC single-stage). But

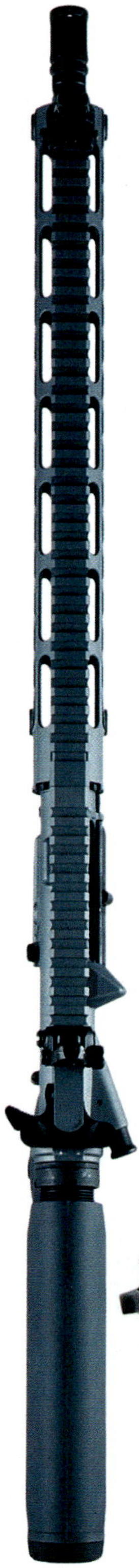

this particular Patrol LW test gun had just returned from a demo trip to a Texas law enforcement agency, where it had been well broken in. After running it like a dog with its ass on fire, this writer is warming up to the QMS as an acceptable everyday trigger.

The Patrol LW doesn't come with any sights on board, so we installed a set of Troy dioptic fixed sights. [Editor's note: The rifle pictured here is fitted with Magpul MBUS Pro Sights.] It seemed a bit counterproductive to pull a sub-6-pound rifle out of the box, and then throw 2 pounds' worth of whizz-bang glass on top. Of course, you could do that and end up with an overall package that is still relatively light, but for this evaluation we stuck to the bare necessities.

The Troy dioptic sights have some nice features. We loved the diamond-shaped rear apertures. They made sight alignment noticeably faster for us than the standard round version. The front sight ramp also has a series of fine horizontal grooves cut into it to reduce glare. While a matter of personal preference, we found the HK-style hood more attractive than the traditional flower-petal protective wings. However, said hood on the front sight gave us a severe case of heartburn because a standard two-sided front sight tool couldn't stretch far enough down to actually adjust the front sight post. By chance, we found a front sight tool that fit (made by Troy, no less) at a local sporting goods store. Buyer beware.

AT THE RANGE

On our first trip to the range, the aforementioned front sight issues precluded any type of accuracy testing, but we were still able to shoot for general first impressions. The rifle is remarkably nimble, compared to this author's personal AR, which weighs in at around 7.8 pounds. Two pounds

TUCSON, ARIZONA
SIONICS
SAR-15
5.56MM

Patrol LW is finished in tungsten Ceracote and looks a million bucks. Note that although the lower is marked with three selector positions, this one is semi auto only. Unfortunately.

doesn't sound like a lot, but it sure as hell feels like it when you start doing things like El Prez drills with long-guns. The other thing of note is that the handguard barely heats up. Why's this even worth mentioning? As previously stated, the MI forearm has an outside diameter of only 1.5 inches, which means that if you happen to place your support hand around the same spot as the gas block, there's less than ¼ inch between your palm and the block itself. We shot the rifle totally bare, save for the iron sights — no rail covers, no gloves. Despite 96-degree heat and more than 100 rounds through the tube on that first day, we didn't notice a thing. The rifle does bounce a little more under recoil than you might be used to, but remember how light it is and that it wears a flash hider. It certainly wasn't a problem, as the 5.56mm round is not particularly famous for its daunting recoil in the first place. Just let it be known, per the laws of physics, if your current AR is set up as a varmint gun, SPR, or open-division contender, the Patrol LW in stock form may seem a touch jumpy to you.

The second trip to the range with the Patrol LW went much smoother. Having adjusted to the change in perceived recoil and dialed in that pesky front sight to a 50-yard zero, we went to work. From 15 to 50 yards, shooting off hand with iron sights resulted in groups averaging about 1½ MOA. A couple groups were sub-MOA, and a couple others stretched out to around 2 MOA. Ammo was all 55-grain loads, a mix of Federal, Hornady, and Tula. A couple guys around the Sionics shop mentioned that they were ringing IPSC-sized steel as far as 600 yards. We didn't witness those feats, but even if you round our groups up to a 2 MOA, it's well within reason.

FAT CAMP CHAMPS

This rifle is just plain fun to shoot. It's handy, points quick, and just wouldn't quit. Even after completing the formal evaluation, we took the Patrol LW out for a third range trip just because it was so enjoyable. The final time, we installed an EOTech XPS-2 and went to town at the 7-yard line. More than once, we neatly clipped the center out of a 1.5-inch circle paster while on the move.

We also shaved several tenths of a second off our average El Presidente times. Even with the rapid-fire drills, there was no overheating of the handguard and no stoppages of any kind — despite having skipped cleaning and lubrication between the range trips. And the last time out consisted of almost 200 rounds of steel-cased TulAmmo, randomly mixed with a handful of Federal bulk.

We also shot the Patrol LW side by side with a S&W M&P 15-22. If you're unfamiliar, the latter is Smith & Wesson's .22 LR AR-pattern rifle. It's a fantastic plinker and trainer, weighing in at a spritely 5.5 pounds. The 15-22 has a quad-rail fore-end, upper receiver, lower receiver, and buffer tube all constructed of polymer. Yet the weight difference between this plastic plinker and the Patrol LW is just 6 ounces. Many folks ascribe to the philosophy of having a .22 LR training weapon for each of their "full power" guns. If you are in this camp, having a 5.56mm AR and a .22 LR AR that nearly balance each other perfectly on the scales will make your rimfire practice an even more effective parallel.

If there's any bad news to be had in this article, here it is: the Patrol LW has an MSRP of $1,299. If you're looking to break into the AR game on a budget, the Patrol LW is not where you'll find entry-level pricing. However, it's hardly an entry-level gun. Premium materials, quality accessories, and an impressively lightweight barrel are assembled in a two-man shop to give you a one-and-done rifle for the whole family. R

Lightweight free float tube has every ounce of fat carved away and keys into the barrel nut to prevent rotation.

ABOUT THE AUTHOR

Tom Marshall works in the private security and corrections industries. He graduated from the U.S. Merchant Marine Academy at Kings Point, where he earned a bachelor's degree, a U.S. Coast Guard third mate's license, and a commission as a U.S. Army officer. During Operation Iraqi Freedom, Tom served as a reconnaissance platoon leader and company executive officer in the last American combat brigade to withdraw from Iraq. His maritime, military, and security experience has taken him to 19 countries on three continents…so far. Born and raised in New York, Tom much prefers his current home in the great American Southwest. There are more guns, more horses, and much less snow.

SIONICS WEAPONS SYSTEMS PATROL LW

CALIBER: 5.56 NATO
MAGAZINE CAPACITY: 30 rounds
BARREL LENGTH: 16 inches
OVERALL LENGTH: 32.75 inches
WEIGHT: 6 pounds with sights
MSRP: $1,300
URL: www.sionicsweaponsystems.com

5 Buildsheet

CONTRACTOR SERVICE RIFLE

Fake Operators, Real Buildsheet

By Tom Marshall

Elsewhere in this issue, you can see the second installment of "Exigent Circumstances" — a short fiction concept we first ran way back in Issue 32. When fleshing out the universe around Hampton-Crane International, our faux private military contracting firm, we realized that even a fictional corporate army would still need a standard-issue rifle. As a team, the RECOIL staff has decades of military and security experience between them. This author holds the dubious honor of logging 13 separate deployments to Afghanistan as a nonfictional government security contractor.

Part of the lure of fiction is that it allows us to live out an idealized or optimized version of our own life experiences. In this vein, we sat down at the builder's bench with one question on our minds: If our fictional protagonists have to carry a company-issued rifle, what would it be? In short, it'd be the "one rifle" we wish we could've taken downrange with us in real life. Thus, the answer will be a little bit different for everybody; we combined our own experience with current trends in DoD small arms improvement programs to come up with this answer, affectionately dubbed the Contractor Service Rifle — or CSR.

The upper half of the Contractor Service Rifle is modeled on the URGI program. The Upper Receiver Group, Improved is a program implemented by Army Special Operations Command to update the M4 carbine with a longer, M-LOK forearm and 14.5-inch mid-length barrel. We used a Rosco Manufacturing Purebred barrel, a match-grade tube made of 416R stainless and capable of ½ MOA groups.

The 14.5-inch Purebreed has a 1:8 twist and 223 Wylde chamber. Both the twist rate and chambering can handle a wide range of both .223 and 5.56mm ammo to allow ammo changes based on mission requirements or the on-hand supply of units being supported by Hampton-Crane contractors. Both the URGI and our CSR sport SureFire Warcomps, which provide a balance of flash reduction and compensation in addition to being compatible with current SOCOM-issue suppressors. Since private contractors have

Though it was built for a fictional story, the Contractor Service Rifle is a weapon we'd gladly take into harm's way in real life.

been known to augment and support Special Operations units, the CSR had to be able to match those capabilities and be cross-compatible with SOCOM standard gear.

For optics, we topped the CSR with a Vortex Razor HD II 1-6x. Variable-power optics allow shooters to use their carbine to its full potential, from CQB to the 500-meter mark. The Razor specifically has been tested and implemented in certain parts of the Special Ops community, and Geissele makes a cantilever mount for this scope that's certified for use by SOCOM, which you also see on the CSR. Other Geissele parts on this build include their Mk 13 rail, Airborne Charging Handle, and Single Stage Precision trigger.

The base receiver set for this build is the American Defense Manufacturing ADM4. The ADM4 lower is a truly ambidextrous lower receiver with both aesthetic and functional improvements over legacy M4 receivers, including ambi safety selector, mag release, and bolt catch. In addition to the Geissele SSP, we used a Primary Weapons System enhanced buffer tube set. The PWS buffer has lengthwise grooves for weight reduction and better clearance of loose debris. The furniture is almost all Magpul, including buttstock, M-LOK covers, vertical grip, mags, and offset iron sights. Magpul products are durable and inexpensive, making them ideal for a corporation looking to outfit hundreds (or thousands) of company-issued guns for hard-duty use. Magpul products have been travelling back and forth to Southwest Asia for a decade at this point, and have held up well in that role.

The entire rifle is finished off in a gen-u-wine rattle can paintjob, skillfully applied by Gabriel Cabrera of Rosco Barrels. This is a personal pattern that Gabe has been working on for a couple years, reflecting his own experience as an independent contractor working overseas. You can find more samples of this pattern under the social media hashtag #CabreraCamo, if it strikes your fancy.

Over the course of modern history, private military contractors have faced steep tactical challenges and prevailed — usually with fewer men and less support than their military counterparts. Executive Outcomes in Sierra Leone, Blackwater in Najaf, and Benghazi are all examples where small numbers of seasoned operators overcame numerically superior forces. While the quality of the men is paramount, cutting-edge individual equipment acts as a force multiplier, allowing contract operators to bring their skill to bear in the quickest, most efficient manner.

The Hampton-Crane CSR is short enough for vehicle and building work while being long enough to engage targets a half-click away. It runs fast and shoots flat, allowing accurate high-volume fire for raid and counter-ambush scenarios. It's suppressor and IR ready for night operations and configurable for shooter preference and mission requirements. It also incorporates real-life lessons learned from several SOCOM-level carbine enhancement programs, including the USASOC URGI program and the NSW Crane analysis of both the M-LOK and mid-length gas systems as superior to legacy carbine configurations. Almost all of the parts used in this build were sourced from companies with current DoD contracts or individual unit-level purchases to reflect commonality between our fictional universe and weapons currently deployed on the battlefield in similar real-life roles. Look for this rifle, and Hampton-Crane International, to appear again in future installments of "Exigent Circumstances." **R**

ITEM		MSRP
Receivers	American Defense Mfg ADM4	$550
Barrel + Muzzle	Rosco/Surefire	$324
Gas System (Block and Tube)	Rosco Mfg	$165
BCG	Rosco Mfg Black Nitride	$134
Handguard	Geissele Mk14 13" MLOK	$250
Trigger	Geissele Single Stage Precision	$240
Light + Mount System	Surefire M600DF/Cloud Defensive/Arisaka	$309
Furniture	Magpul MOE	$99
Pistol Grip	VZ Grips	$95
Safety	Ascend Armory Short Throw	$40
Optic	Vortex Razor HD II 1-6x	$1,999
Mount	Geissele Super Precision SOPMOD	$325
Totals		**$4,530**

7.62X39 AR

We Build an AR That Takes AK Mags

By Steven Kuo

In this issue of RECOIL, we celebrate the AK, in all its splendor and variations. What better way to honor it than by building a 7.62x39mm AR. And not one that takes weird unreliable mags shaped like the letter J, but one that happily feeds from the original banana mag.

Billet Rifle Systems offers a receiver set tailor made for this build. The lower receiver has a yawning AK magwell, a stainless steel cross pin, and a paddle mag release to accept rock-n-lock mags like an AK-inspired AR truly should. The upper sports a sweet non-reciprocating side charging handle, on the left side where proper-handed shooters expect it to be. BRS sells entire rifles too, but we wanted to roll our own AK tribute, so we used their upper, lower, action spring, and custom bolt carrier group as the basis for our build. Unlike CMMG's 7.62x39 Mutant, with its stretched dimensions, BRS's receivers conform to the standard AR-15 footprint — thus explaining the slightly cramped trigger guard.

Our next selection was Faxon's nitrided 16-inch barrel with mid-length gas system. BRS recommended opening up the gas port by a few thousandths of an inch for optimal function. We topped the barrel off with a SureFire muzzle brake and suppressor adapter so we could attach a can for our next hog hunt.

An under-folding stock wasn't an option, but we put a side-folding stock on it with LAW Tactical's clever stock adapter. Mission First Tactical's buttstock was both functional and evocative of metal folding stocks found on various AK-pattern rifles. You can't discharge the rifle with the stock folded, but it's very handy for transport — and, let's be honest, it looks cool and was perfect for our AK theme.

When you picture an AK in your mind, we'd bet that it has wood furniture. It took some digging, but we found hand-finished walnut rail panels and pistol grips from Black Wood USA; their military red finish was perfect for our theme. The M-LOK panels are a little chunky, a great match with Centurion Arms' very low-profile 15-inch handguard.

On top, we attached a Burris XTR II 1-5x scope with CQ mil reticle, a nice match for this rifle. We initially put it in a QD mount but kept hitting the levers when working the side-charging handle; we switched to an Aero Precision mount, completely smooth on the left side and very light and affordable to boot. For backup iron sights, we mounted Sidewinder sights from Strike Industries, which flip out as 45-degree offset sights or up as traditional sights.

The other parts, pins, and springs to complete the build we also got from Strike, except for the ambidextrous safety selector from AIM Surplus. We also installed a drop-in trigger but found it wouldn't ignite the hard primers in 7.62x39mm ammo roughly 5 percent of the time. We swapped in a crisp HIPERFIRE Eclipse trigger, setting up its springs to deliver more hammer energy

than stock, and the problems went away.

Finally, we always install a light and sling on rifles intended for use beyond competition. The SureFire M600DF scout light puts out 1,500 blinding lumens with its rechargeable 18650 battery, mounted in a slick Impact Weapon Components Thorntail2 mount, which pulls the light in tight against the handguard. A suitably austere but functional Proctor sling with QD swivels finished off the build.

At the range, we tried eight different types of ammo, both domestic and imported. All ran perfectly, aside from the light strikes before we switched triggers. To assess accuracy potential, we benched it with a Nightforce at 40x — and it delivered 1 MOA five-shot groups with Federal's Power-shok (the most accurate load in our 7.62x39 ammo roundup in Issue 26) and Winchester's Super-X ammo. Who says 7.62x39mm can't be precise? Other groups generally ranged from 1.5 to 2.5 MOA, and most loads posted average muzzle velocities just over 2,300 fps on our Magnetospeed chronograph (at 92 degrees F and 32-percent humidity). As usual, the Magnetospeed proved its worth, as we strapped it to the handguard and left it there during accuracy testing while simultaneously capturing velocity data.

One item to note: BRS's receiver is tuned specifically for Magpul's MOE AK PMAGs, which ran 100-percent in our testing. Magpul's Gen M3 magazines surprisingly have slightly different geometry and wouldn't quite clear the mag catch to lock in place. Some steel AK magazines would lock in, but experienced some feeding problems. Bottom line — use MOE PMAGs with this gun, and it won't let you down. Also, the bolt doesn't lock back when empty, but neither do AKs.

Weighing in at 9.3 pounds unloaded, the rifle handled beautifully and shot softer and flatter than we expected for a 7.62x39. BRS spent a lot of effort dialing in their system, and it shows. With the Eclipse trigger set at 3 pounds, we ripped off smoking splits with huge grins on our faces. This gun is a blast — we're definitely taking it to some rifle matches. It turned out great, and we hope Mikhail would approve. R

ITEM	MSRP
Billet Rifle Systems BRS-47 Receiver Set (Side Charging)	$649
Billet Rifle Systems 7.62x39mm Bolt Carrier Group	200
Billet Rifle Systems Action Spring	20
Faxon Firearms 7.62x39mm 16-inch Barrel	185
Faxon Firearms Gas Block and Tube	46
SureFire 7.62 Muzzle Brake	149
Centurion Arms M-LOK CMR Handguard	240
Black Wood USA Wood Rail Panels and Handstop	115
Black Wood USA Featherweight Wood Grip	150
LAW Tactical AR Folding Stock Adapter	269
BCM Carbine Stock Hardware Mounting Kit	65
Mission First Tactical Battlelink Minimalist Stock	60
HIPERFIRE Hipertouch Eclipse Trigger	275
Strike Industries Lower Parts Kit	35
Strike Industries Enhanced Ultimate Dust Cover	16
Strike Industries Forward Assist	19
AIM Surplus AR Ambi Safety Selector	23
Burris XTR II 1-5x24mm Riflescope	959
Aero Precision Ultralight SPR Scope Mount	95
Strike Industries Sidewinder Backup Iron Sights	160
SureFire M600DF Scout Light	299
Impact Weapon Components Thorntail2 Light Mount	50
Impact Weapon Components QD Sling Mount	19
Proctor Sling w/ QD Swivels	44
Magpul PMAG 30 AK/AKM MOE	14
Totals	**$4,156**

Build Concept: A defensive build based on the newly available Radian Weapons receiver that serves as a testbed for the Dracos Straightjacket barrel and Cerakote's new Elite coatings.

By Rob Curtis

RADIAN DRACOS CUSTOM

Radian Weapons recently got back to producing its AX556, featuring the ambidextrous A-DAC control system. We were fans of the setup back when Radian was known as AXTS, having shot a few AX556-based builds years ago. We added it to our wish list, but as the company rebranded and focused on building complete rifles, our chances to get one evaporated, only intensifying our desire to own one.

Fast-forward a year and imagine our excitement as Radian catches up with rifle demand and releases full receiver sets. Sure, they're spendy, but our lust was fully fueled by all those discussions board convos about them. And, when it comes to cost, let's just say we always knew an AX556 build wasn't gonna be a budget build.

We'd also heard about a little company called Teludyne and its Straightjacket process. They'd take your barrel, turn it down to its barest outer diameter shove it in a tube and fill the gap with goo. The result, according to Teludyne, is a longer-lasting, lighter barrel with reduced harmonics and an immunity to thermal loading that increases its accuracy. As neat as the concept was, we didn't have the patience to wait on a barrel conversion, or deal with the company if something went wrong. But when Falkor Defense bought in and made its line of Dracos barrels using the Straightjacket tech (and guaranteeing the barrel for life), we figured it'd be the perfect way to offset the AX556's beefy barrel nut/handguard interface. And was it.

With our setup, the rifle's balance point is exactly at the base of the handguard/upper junction. We accessorized the rifle with Aimpoint's new CompM5 so we'd never have to worry about breaking our optic or changing the battery. We'll skip the BUIS with an optic as tough as the CompM5, but we'd never run a non-competition carbine without a light. We got our hands on a prototype SureFire M600DF light that puts out 1,500 lumens for more than an hour using an 18650 rechargeable lithium-ion battery.

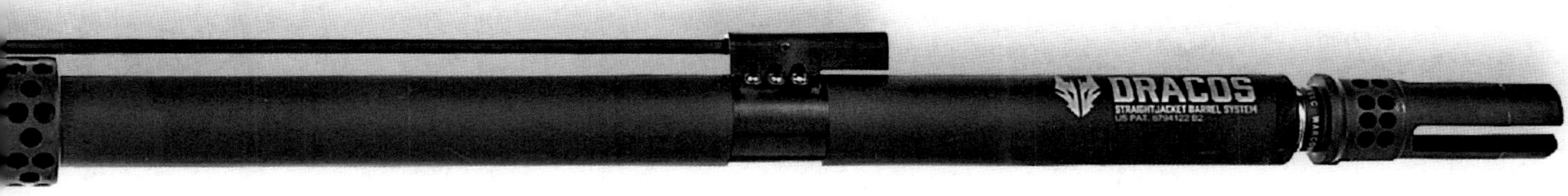

We're stoked to see SureFire get off its watch ight kick and build *exactly* what we've been begging for!

The coating is Cerakote's new Elite series. We've used it on a few components, but we'd yet to see applied in multiple layers. We know Radian holds ridiculously tight machining tolerances, so tight that they account for the thickness of the standard Cerakote used on all their parts. They don't normally do patterns on their guns because the added layers of Cerakote can make things too tight. But we worked with Radian and Blown Deadline Custom Cerakote to measure the difference and found the new coating had no effect on fit, even built up in layers needed for the MultiCam Black effect, above.

Note, we tested the AR Gold in our trigger guide and noted similar findings. The company's claims about short reset are true. This trigger is incredibly fast, clean, and reliable, but the trade-off is the absence of tactile reset.

As shown, the rifle shot five five-round groups using Federal Premium 223 Rem 77-grain GMM that all printed better than 0.65 MOA. The best was 0.52 MOA. Recoil was negligible, and we haven't had any stoppages. Though, we're still shy of 1,000 rounds. **R**

ACCESSORIES	MSRP
Blown Deadline, Cerakote Elite MultiCam Black	$440
Magpul M-LOK Offset Light/Optic Mount	35
Aimpoint CompM5 and AR Mount	1068
SureFire M600DF Scout Light	299
SureFire 18650 Li-Ion Battery	50
Magpul M-LOK Rail Cover, Type 1	16
Magpul M-LOK Aluminum Rail, 5 Slots	20
Totals	**$1,928**

COMPONENETS	MSRP
Dracos StraightJacket Barrel, 16-inch 223 Wylde	$850
Radian Weapons Enhanced BCG	150
JP Custom Operating Spring OSC	20
Brownells H2 Carbine Buffer	30
Radian Weapons Raptor SD	99
Magpul MOE Grip	20
Magpul M-LOK Rail Cover Type 1	16
Magpul PMAG 30 GEN M3	15
Surefire WarComp-556, 1/2-28	149
Radian Weapons Ultralight QD Endplate	26
Brownells AR Receiver Extension	26
Radian Weapons AX556	370
Radian Weapons 15.5-inch	600
Radian Weapons Talon Ambidextrous Safety	50
Magpul MOE SL-S	80
Radian Weapons takedown pins	26
American Trigger Corporation AR Gold	279
Totals	**$2,790**

AN MK18-ISH FOLDER

By Rob Curtis

The Mk18 is as iconic as it gets when it comes to SBRs. This is not an Mk18. This is what we think the Mk18's punk-ass little half-brother would turn out like if its NSWC Crane papa stepped out and knocked up a young mistress.

The Mk18's developmental roots go back to the 1960s and the original shorty AR, the Colt Commando.

NSWC Crane took inspiration from the Colt Commando when it made its CQBR drop-in upper receivers for U.S. Special Operations forces. These were 10.3-inch Colt-barreled uppers that eventually became the Mk18 when combined with surplus M16 rifle lowers from the Navy's stocks. One of the most distinctive parts of the Mk18 is the flat dark earth RIS handguard. This was originally a Knight's part, but Crane would later shift to a handguard made by Daniel Defense.

The Mk18 story gets confusing at this point because DD capitalized on the Mk18 name when it won the contract to produce the handguard. It wrapped its own parts in the Mk18 handguard, christened it the "Mk18 Upper Receiver Group," and made it available on the civilian market.

The DD Mk18 upper, which is the basis of this issue's Buildsheet rifle, is basically a DD's recreation of the Colt-produced Mk18.

We say *basically* because, though the DD handguard is an authentic SOPMOD (i.e. government issued) part, the upper is an Mk18 in name only. DD used its own cold hammer forged 1:7 twist 10.3-inch barrel and upper receiver in place of the Colt parts found on the real deal.

But we weren't trying to build a true Mk18, just a 10.3-inch AR, so we don't have to nitpick all the differences between the DoD Mk18 and the DD version of the Mk18. We wanted a versatile, fast-handling rifle that's easy to stow out of sight while being quick to the fight, whether serving as a truck-gun, CQB platform, or a home-defense tool.

Our shorty is set up for CQB. The heart of the Mk18ish is Law Tactical's ARIC abbreviated bolt carrier group, giving our AR15-based carbine a fire-when-folded superpower when combined with Law's folding stock adapter. Honestly, we got the ARIC and wanted to build a rifle around it, and this setup made a lot of sense.

“WE WANTED A VERSATILE, FAST-HANDLING RIFLE THAT’S EASY TO STOW OUT OF SIGHT **WHILE BEING QUICK TO THE FIGHT.”**

We started off with our trusty lower Mega Arms billet receiver. We chose it, first, because Mega made a tightly toleranced, ambi lower, and second, because it was already Form 1’d for use as an SBR. The Geissele SDE-E trigger might be overkill for a 1 MOA gun, but it’s fast, consistent, and above all, reliable. Reptilia’s CQB grip provides an upright grip angle that’s more comfortable and efficient for rifles with short (or folded up) buttstocks. And out back, we’ve got Magpul’s Gen 2 UBR stock. It’s heavy, but it offsets the weight of the can, light, and laser we’re running up front and places the rifles point of balance right at the magwell.

We had to force the UBR to get along with the folding stock adapter using a Dremel. It took 10 seconds to radius the leading edge of the hinge-side, polymer stock cover to create clearance for the stock to fold flat against the receiver.

Up front, we added an Insight ATPIAL IR laser and a SureFire light that’s IR and white-light selectable. Both light and laser are activated with a rail-mounted, dual-button Unity TAPS switch. The nightfighter theme continues with the EOTech XPS3. Its lowest reticle settings are compatible with night vision tubes, if you want to mount a night vision monocular behind it. It’s also sitting up on a riser for a more efficient, heads-up shooting posture.

ON THE RANGE

The SBR is well-mannered thanks to the recoil-absorbing, SureFire SOCOM 2 556 QD suppressor. Splits are fast and the muzzle confidently swings between targets without over-running them during transitions since the gun is well-balanced.

Accuracy is what we expect from a 10.3 running XM193 and M855, just over an inch. Though groups tightened up to just under an MOA with better ammo. Around 300 rounds in, we began experience cycling issues with the can puking carbon back into the BCG. The short, light ARIC bolt carrier didn’t have the inertia to overcome the friction provided by the combustion debris blown back into the upper receiver and slowed to the point of failing to go into battery on several shots. We lathered the BCG in CLP, and the gun was GTG till COB that day.

The gun ran unabated, whether the stock was folded or open. As sexy as the capability sounds, firing a folded-up AR should be reserved as a last-ditch, we’re-getting-ambushed sort of option.

Overall, the build is a ton of fun to shoot, both in daylight and in the night. Don a PVS-14 and use the ATPIAL to poke holes in unsuspecting things a block or more away. Chin-welded daylight (or white light) engagements with the big, bright EOTech riding high is a revelation of comfort, speed, and situational awareness when you’re used to a head-down, cheek-mashed posture on the gun. **R**

PARTS	MSRP
Daniel Defense Mk18 Complete URG	$1,242
Law Tactical ARIC Bolt Carrier System	$369
Mega Arms GTR-3S Billet AR-15 Lower Receiver (Discontinued)	$200
Magpul UBR GEN2 stock	$200
Law Tactical AR Folding Stock Adapter Gen 3-M	$270
Geissele Automatics - Super Dynamic Enhanced (SD-E) Trigger	$240
Battle Arms Development BAD-ASS Short Throw Safety	$85
Battle Arms Development Enhanced Pin Set	$30
Reptilia CQG Pistol Grip	$20
Radian Weapons Raptor LT Charging Handle	$65
SureFire WARCOMP-556-CTN-1/2-28	$149
SureFire SOCOM556-RC2	$1,099
EOTech XPS3	$655
Kinetic Development Group SIDELOK Universal Riser	$90
Dueck Defense RTS Fiber Optic Iron Sights	$250
Unity Tactical TAPS	$152
SureFire M300V Scout Light	$429
Insight ATPIAL (LA-5B/PEQ)	$1,440
Total:	**$6,985**

M16A4

OIF Edition

By Dave Merrill

It's no secret the firearms market has softened since Trump took office. Prices are lower than they've ever been, so we're seeing a lot of interest in just-for-fun guns, and clone builds in particular.

The author on a fun street in Fallujah, November 2004.

With a nod to the OIF vets among you, we're looking at an M16A4 clone; specifically, a clone from Operation Iraqi Freedom, mostly from the author's service rifle, circa 2004.

Shortly after the initial invasion of Iraq, the USMC began replacing well-worn M16A2 rifles with the M16A4. Though it still sported a fixed stock and a 20-inch barrel (the better to do rifle drill with), it had a flat-top upper receiver that simplified the mounting of optics and a Knight's Armament M5 Rail Adapter System (RAS) to hold lights and other accessories.

The clone's complete upper comes from Aero Precision. It's made for their M16A4 Special Edition Rifle. The handguard is a P&S Products M5 RAS, an acceptable-and-issued substitution for the original Knight's Armament handguard, which is harder to find. There's nothing special about the lower receiver, A2 stock, and internals. We plucked these parts from the RECOIL workshop's spare parts bins.

The first Trijicon ACOGs issued with these rifles were the TA31F variety, later to be replaced with the TA31RCO-A4 with LaRue mounts. Shown is a TA31F with a LaRue QD LT100 mount and TA91 Tenebrex killflash complete with the Trijicon objective flip cap.

It's the small things that make this rifle just right. It's also the small things that make this rifle look a bit ridiculous. Take the carry handle attached to the rail, for instance. It's unclear whether it was kept on the

rifle as a backup sighting system or to keep the part from getting lost. The author still has no idea, and he did this on his own issued rifle.

The buttstock magazine pouch was in common use in Iraq. While on a forward operating base and not outside the wire, weapons were carried empty or, in Marine parlance, kept in "Condition 4." The buttstock pouch was an easy way to keep a magazine with the rifle without it being loaded. The buttstock magazine pouch also provided a sling location that was actually worth a damn.

Rifle slings were in a state of flux at the time this rifle was in service. The complicated three-point sling was the original issued item with the M16A4. Many Marines either found more useful two-point alternatives, or modified the complex three-points. In later years, Viking Tactics VTAC slings as you see here, were standard.

Both the SureFire M962 (seen in the lower-right photo), as well as the two-cell SureFire M951 (shown with a body extension converting it into a three-cell in top photo) are period correct. Back in the days when 65-lumen lights were the standard, 200 lumen lights like these (even with a 20-minute run time) may as well have been the light of God.

Other items like the magazine coupler and worn-as-hell USGI mags, hideous in hindsight, came directly from a seabag that sat untouched in the author's garage for more than a decade.

Though this rifle won't be humped through the streets or foreign hovels, it will still ping steel at 600 yards. Holding the A4 resurrected from the author's OIF service had him falling in love all over again. **R**

ITEM	MSRP	STREET/USED
Aero Precision A4 Upper	$644	$410
Complete Lower w/ A2 Stock Assy	400	250
P&S M5 RAS	276	125
KAC M5 Panel Kit w/ VFG	210	50
Trijicon TA31F ACOG	1,502	900
Trijicon TA91 Killflash	48	41
Trijicon Objective Flip Cover	18	15
Colt Carry Handle	130	55
SureFire M962 (kit w/ WML)	1,100	400
Viking Tactics VTAC Sling	50	40
Spec-Ops Mag Pouch	30	10
Totals	**$4,408**	**$2,296**

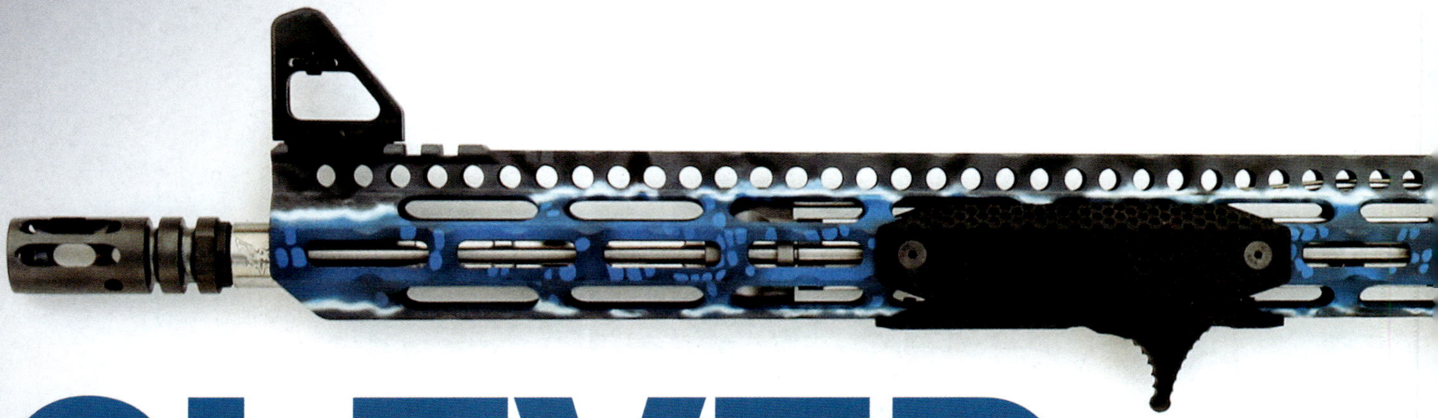

CLEVER GIRL

A Different Kind of Thin Blue Line Build

By Tom Marshall

We try to make sure that every entry in the Buildsheet column has a purpose or desired end use that appropriately shapes the build itself.

In this case, the idea came to us while coaching a new shooter with little-to-no experience running carbines in a dynamic training environment. So what *would* be an ideal rifle for someone not accustomed to shooting, or moving with 3 feet of pew pew in their hands? Lightweight would be highly desirable so it could be wielded by students of all sizes and builds. Adding features to mitigate recoil could also be useful.

Yes, yes, we know. Software over hardware, good training over cool gear. And we agree. But if we can utilize hardware advances to shorten the learning curve and more effectively suss out needed corrections, why pass those up just to thump chest? Accuracy ought to be as good as it possibly can, at least from a mechanical standpoint. This isn't to "make the shooter better," but to remove all doubt that misses and fliers are technique based, not the result of cutting corners on the build.

To shave weight, we talked to two companies who've made very good names for themselves in the light-and-fast arena of AR builders: 2A Armament and Battle Arms Development. The foundation of this gun is 2A's Aethon receiver set, which weighs in at just over 13 ounces stripped. In the interest of keeping this rifle svelte, they also provided titanium takedown pins, gas block, and T4 compensator as well as a fluted buffer tube. Battle Arms Development came with their "BADASS" short-throw ambi safety selector, enhanced and skeletonized bolt catch, and enhanced mag release. The RACK charging handle you see here is also of Battle Arms provenance.

One of the most unique features on this build is the barrel, which comes to us by way of Northtech Defense. Their new Synergy barrels are a hybrid of titanium and 416R stainless. The obvious goal is the accuracy of stainless with the weight savings of titanium. Their 16-inch mid-length gas barrel weighs in at 22 ounces.

We sleeved the Northtech barrel with a Midwest Industries G3 M-LOK handguard. This forend is of the growing ilk that deletes most of the top rail for added M-LOK slots to reduce weight and enhance ergonomics for tape switches and light mounts. There's still a short section of Picatinny rail at the muzzle end for a front iron sight. We added a set of fixed irons from Scalarworks. These provide enhanced functionality and reduced weight over legacy irons. We also utilized a Scalarworks mount for the Aimpoint Comp M5. A simple, rugged, no-nonsense optic, the M5 stands in a lower 1/3 co-witness.

The guts of this gun were given equal consideration. The bang switch is a KE Arms SLT-2. What we really love about this trigger, from a teaching standpoint is that KE's unique Sear Link Technology

allows you to press the trigger and receive a positive click even with the hammer down. It also allows you to put the selector switch on Safe with the hammer down. This gives you a semi-automatic dry-fire capability, and full manipulation of the selector lever, without having to run the charging handle at all.

Once it's time to go live, the Iron City Rifleworks G2 bolt carrier group comes into play. This group features a carrier, skeletonized at the rear for lighter weight and faster cycle time, matched to a bolt of 158 carpenter steel — both coated in their black diamond finish. We paired it with a Sprinco white buffer spring and a CTS AKTIVE buffer. Tuning an AR's recoil system is a bit of an art. Combining buffers, springs, gas blocks, gas tube lengths, and muzzle devices requires some trial and error. We have experience with all of the components listed here and have gotten them to run very effectively in this or similar combinations. Your mileage may vary.

A VZ Grips pistol grip and a set of Rail Scales XOS panels accomplish front and rear grippage. The XOS panels provide full-wrap coverage with the Midwest handguard and can be had with cut-outs for seamless integration of vertical grips and hand stops, as pictured here.

Last, but most certainly not least, we sent the whole package off to P4 Coatings for the namesake GunKote job. For the uninitiated, this color scheme was hand-matched to the velociraptor "Blue" featured in the movie *Jurassic World* — to include dino-scale texturing. We think Jason at P4 slam-dunked fidelity to this prehistoric heroine, and tip our hats to our newly minted rifleman (rifle-*woman*) who came up with the concept art.

The end result is a highly capable carbine that tips the scales at 5.62 pounds exactly as pictured and provides a new shooter with a platform that's easy to handle, with a lot of built-in room for growth, that's sure to turn heads and ring steel in equal measure. **R**

PARTS	MSRP
Aethon Receiver Set	$500
Blue Force Gear Vickers push-button sling	$80
Handguard: Midwest G3 ML 15"	$200
Charging handle: BAD RACK	$125
Barrel: Northtech Defense Titanium	$689
Optics Mount: ScalarWorks Lower 1/3	$149
Optic: Aimpoint CompM5	$800
Fixed Irons: ScalarWorks	$239
Grip/covers: Rail Scales XOS	$233
BCG: Iron City Rifleworks Lightweight DLC	$330
Titanium Gas Block	$70
T4 Titanium Comp	$130
Mag release: BAD EMR	$22
Bolt catch: BAD Enhanced	$15
Stock: MFT Minimalist	$60
Grip: VZ Recon Gen 2 Short	$85
Selector: BAD Ambi Short Throw	$60
Trigger: KE Arms SLT-2	$199
Buffer: CTS Active	$68
Buffer Tube: 2A Flute	$60
Pins: 2A Armament Titanium	$29
Custom Coating: P4 Coatings	$650
Total:	**$4,793**

THE RIFLE THAT wish BUILT

We Build a Gun With Parts From the Worst Vendor on the Internet

By Dave Merrill

Over the years, we've seen all sorts of crazy cheap deals on parts and pieces; some of them with questionable origins. The perfect place to put them all together was right here in our budget issue. Who needs Black Friday when you can build yourself a working rifle from the cheesiest of Chinesium from Wish.com — or at least try to.

We did our best to avoid products that were simply stolen intellectual property from American companies. Despite our best efforts, the iron sights we purchased ended up being garbage knockoffs of Magpul MBUS sights — what was pictured in the listing was totally different. Still, they ended up being absolutely worthless due to two reasons: The apertures freely rotated, and the handguard was total trash.

PARTS AND PIECES

When the handguard arrived, it actually looked pretty decent. Not only was it lacking that chalkboard-made-in-China feel, it had anti-rotational tabs, a barrel nut made of steel, and a whopping five quick-detach sling points. If you put it on a shelf in a gun shop, no one would believe it only cost $27. Ultimately, it ended up being LG;FU. That is to say: Looks Good; F*cked Up.

None of the QD points were in-spec; all were undersized. When the barrel nut was properly aligned to allow the gas tube to enter the upper receiver, the rail was canted several degrees. We had to shave down the alignment tabs to line up the rail with the upper. The handguard itself was a study in poor aluminum extrusion; rather than remaining in-line with the upper from tip to tail, it nearly touched the barrel at the muzzle end. Even if the Magpul-knockoff iron sights were OK, the rail itself precluded their use.

One of the few components we couldn't procure from Wish was the barrel. No matter, there are several über-budget options. We went with a Bear Creek Armory blemished barrel. We know what you're thinking, but no, not

all of them are blems, apparently. This mid-length 5.56mm barrel had a gas port mic'd at 0.088 inch — too large to start out with.

The lower receiver was an Anderson Arms purchased on sale for $25, and the upper receiver a blem from Gorilla Machining with the original intended manufacturer's markings milled off of the side. We skipped the ejection port cover, because we didn't want to spend the extra $2.

The gas block, shockingly, was made from steel. Instead of a press-on, it's the far inferior clamp-on type that went out of favor more than a decade ago because of gas leaks. The finish came off when you looked at it.

Though sold on Wish, the listing for the bolt carrier group claimed it was U.S.-made from a company called KM Tactical. A cursory web search brought up a company based in Missouri with an online store selling a lot of similar items. As expected, it wasn't marked as high-pressure-tested — not that we'd believe it if it were.

The lower parts kit was a mere $20, and the included springs looked like it. Surprisingly, it also included a winter trigger guard and an ambidextrous safety selector. The only part we were missing for this build was a pistol grip, so we pulled a black Hogue from a parts box.

OUTFITTING

Since the iron sights were out, we decided not to buy a complete fake on Wish and turned to Vortex for an inexpensive red-dot sight. We chose a SPARC AR that runs on a single AAA battery. This would ultimately be the highest quality item that would ever touch this rifle. While we believe a weapon mounted light to be paramount for any gun configured for defensive purposes, we skipped it here because we couldn't imagine a single scenario where we'd stake our lives on this rifle.

It's not lost on us that the MSRP of the Vortex SPARC AR exceeds the rest of the rifle combined.

AT THE RANGE

The rifle experienced a 50-percent failure rate right from the start — frankly, better than we expected. The main culprit was that the trigger wasn't resetting. Under normal circumstances you'd check for incorrect hammer spring installation, but in this case, it was because the springs were made of zinc or butter.

COMPONENTS	MSRP
Anderson Manufacturing lower receiver	$25
Gorilla Machining blemished upper	$25
Bear Creek Arsenal blemished barrel	$45
Complete buttstock assembly	$24
Gas block and gas tube	$14
KM Tactical NiB bolt carrier group	$68
Handguard	$27
Charging handle	$8
Lower parts kit	$20
Total:	**$257**
ADDITIONAL COMPONENTS	
Vortex SPARC AR	$259
Black Aces 5.56 Po' Boy Silencer	$199
Price as configured:	**$715**

Once that was corrected, the rifle actually ran. As to how long it'll continue to run? Experience tells us it won't hold a candle to a properly built, in-spec rifle. But we'll continue to flog it and chronicle the misadventures on RECOILweb. Because we're masochists.

LOOSE ROUNDS

The fact is, a cheap build may not end up being as cheap as you thought. Firstly, none of the prices reflected in this article showcase how much shipping actually costs, nor how much you'll wait for your wares. Additionally, a handful of the subpar items we received had to be replaced entirely, adding

to the overall time and cost. Furthermore, we received parts that didn't match the descriptions, and we have a handful of well-known failure points in this rifle as well.

We did give this rifle the worst paint job we can think of. So, there's that.

Ultimately, it was an interesting thought experiment, but what we received wasn't all gravy. You'd be better off looking for a sale on an inexpensive AR, such as a Palmetto State Armory, rather than trying to piece everything together from a sketchy foreign vendor. Unless you just want to make your partner cry and your gunsmith rich. R

"YOU'D BE BETTER OFF LOOKING FOR A SALE ON AN INEXPENSIVE AR, SUCH AS A PALMETTO STATE ARMORY, RATHER THAN TRYING TO PIECE EVERYTHING TOGETHER FROM A SKETCHY FOREIGN VENDOR."

HAWAII FIVE-FIVE-SIX

Building a rifle should be fun. Yes, a rifle is a tool, but it can also reflect your personality and be a form of artistic expression. Furthermore, it's not as hard to assemble a rifle as you might think.

In fact, Aero Precision has a monthly offering that does exactly that — builder sets that combine straightforward assembly with artistic designs. These sets include a lower, upper, and handguard that have been Cerakoted in different themes by various artisans. In months past, for instance, Weapon Works, Blown Deadline, Republic Rifle, Nevada Cerakote, MAD Custom Coating, and others have created all sorts of designs, from various cool colors to all types of camouflage patterns to aviation livery to Veterans Day themes.

The components themselves have convenient features designed to make assembly easier, including threaded pins that eliminate the frustration of installing the bolt catch and springs launched into the nether regions of your garage, as well as handguards that align easily and clamp down evenly. The Aero M4E1 builder set forming the foundation of the build in this issue is a custom design that we were fortunate to snag from them, painted by MAD Custom Coating in a unique Hawaiian theme. The receivers are forged, yet have numerous nice additional touches like the integrated trigger guard and flared magwell that you might expect in heavier, more expensive billet sets. The handguard is Aero's Atlas R-ONE 15-inch M-LOK, providing plenty of real estate for the support hand and accessories.

We chose Ballistic Advantage's bead-blasted 16-inch .223 Wylde barrel, which was spun from 416R stainless steel with a BA Hanson profile and came fitted with a pinned low-profile gas block. We capped it off with a VG6 Gamma 556 muzzle brake. The shooter's interface with the gun is always important, so we outfitted the rifle with a Rise Armament RA-535 trigger and comfy Magpul furniture. The trigger broke at a feathery and crisp 2.75 pounds, with a short reset.

Even though the paint scheme is light-hearted, this rifle can still do serious work. And for that, a light and slings are must-haves. SureFire's scout lights are among our favorites for rifles, their brand-new Scout Light Pro Dual Fuel variant sporting a built-in, adjustable M-LOK mount. There are several great light mounts for scout lights on the market, but the new Pro is about as low profile and minimalistic as you can get. It places the light closer to the handguard than a creeper on the New York subway. The sling is Blue Force Gear's popular Vickers two-point quick-detach sling.

EOTech's Vudu 1-6x SR1 scope topped off the rifle, providing crisp glass and a first focal plane reticle with a circle-dot at 1x and a precise Mil scale at 6x. It worked well at either end of the magnification range, though illumination is just barely daylight bright — more like a slight reddish hue in bright sunlight. A handy lever screws into

Getting Artistic With Our Aloha Aero Build

By Steven Kuo

SPECS

CALIBER:
.223 Wylde

BARREL LENGTH:
16 inches

OVERALL LENGTH:
33 to 36.4 inches

WEIGHT UNLOADED:
9.4 pounds

PRICE AS FEATURED:
$3,977

COMPONENT	MAKE	MODEL	MSRP
Lower, upper, handguard builder set	Aero Precision	M4E1 builder set with 15-in ATLAS R-ONE M-Lok handguard	$550
Barrel	Ballistic Advantage	16-inch .223 Wylde BA Hanson premium series stainless steel barrel with mid-length gas and lo-pro gas block	$245
Muzzle device	VG6	Gamma 556	$80
Gas tube, pin	Aero Precision	Melonite gas tube	$17
Bolt carrier group	SureFire	Optimized bolt carrier group	$399
Charging handle	Aero Precision	AR-15/M4 5.56 ambidextrous charging handle	$80
Receiver extension, buffer, spring, castle nut, end plate	Aero Precision	AR-15 carbine buffer kit	$40
Buttstock	Magpul	MOE SL-S carbine stock	$80
LPK with pistol grip, less FCG	Aero Precision	M4E1 MOE lower parts kit minus fire control group	$45
Fire control group	Rise Armament	RA-535 trigger	$259
Optic	EOTech	Vudu 1-6x precision rifle scope SR-1	$1,399
Scope mount	EOTech	PRS QD cantilever ring mount	$195
Iron sights	Magpul	MBUS Pro	$190
Sling	Blue Force Gear	Vickers padded sling	$60
Sling mount	Magpul	M-LOK QD sling mount	$20
Light	SureFire	Scout Light Pro Dual Fuel	$319
TOTAL			**$3,978**

the zoom ring, making it easy to dial in your desired magnification, but it's too long for our liking, so we'll probably shave it down.

We thoroughly shook out the rifle at an urban precision rifle course at Thunder Ranch, under the tutelage of none other than the legendary Clint Smith. We engaged paper and steel targets at 5 to 500 yards from various shooting positions and in challenging courses of fire, all while meeting stringent marksmanship requirements. We practiced movement, use of cover, malfunction clearance, one-handed manipulations, and surgical shots. Clint continually hammered us on mindset, and the critical importance of always thinking logically. We fired around 700 rounds of Doubletap ammo through our rifle without incident, and during some Oregon downpours. The rifle handled wonderfully, with great ergonomics and minimal recoil impulse.

After our Thunder Ranch outing, we received SureFire's new optimized bolt carrier group. Since we plan to run this rifle with a can in the future, as all rifles should be, we swapped out Aero's nitrided BCG to test the SureFire unit. Thus far, it's run smoothly and perfectly while unsuppressed.

On the bench, we discovered the 1:8 twist BA barrel really loved heavy bullets and swiped left on the light ones. 75-grain Black Hills match traveling at 2,531 fps printed 1 MOA five-shot groups, while plebeian 55-grain loads spread out to 3 MOA. Gorilla 69-grain SMKs slotted in the middle at 1.4 MOA. So, whether hosing or placing precision shots, our build gets it done — and in style.

Hopefully, we've inspired you to get artistic and have some fun with your next build. Whether you paint it yourself (see our DIY in Issue 13) or use an out-of-the-box builder set like those offered by Aero Precision, make yourself a piece of art! Aloha! **R**

COYOTE CRUSHER

A Lightweight 22 Nosler Folder that Turns Predator into Prey

By Rob Curtis

There's got to be a name for it. That thing that happens when you get something you're not really excited about, but it becomes one of your favorites over time. It could be a tool you bought on a whim, and that lack of emotional connection leads to some pretty hard use that then leads to endearment, as long as the thing performs.

Well, as gun people do, we picked up a 22 Nosler barrel when it was the hot new round. That barrel sat on a shelf, dutifully waiting for its time. It idled, witnessing the birth of its nemesis, the 224 Valkyrie cartridge. Struggling through a few complications in its infancy, .224 got over the hump and is on its way to maturity.

22 Nosler, though, seems stuck in a permanent awkward teenager phase. It's extremely fast and has a long reach, but it's a little on the hormonal side. It's taken to bullying and stealing its rebated-rim brass' lunch money.

The cartridge may be a little temperamental, but on the whole, our 22 Nosler build went from socks-for-Christmas to front-of-the-safe problem solver without our realizing it.

Sure, 22 Nosler is tough on brass, but we're not reloading it, nor are we shooting enough that the cost of factory ammo prevents us from reaching for it when the coyotes start yipping.

GOALS

The overall goal of this build was to make a handy AR that could ride around a farm in a truck or get carried into the fields where it could cut down unwelcome predators. To that end, three key components were selected. The Proof Research 22-inch Camgas carbon-fiber composite barrel wrings every last bit of velocity from the 22 Nosler, and it's light for its length. A Law Tactical folding stock adapter offsets the barrel's length. And the added weight of the folder and long barrel is further offset by a lightweight titanium bolt carrier from Brownells.

Up front, we opted for Geissele's MK18 handguard due its full-length 1.5-inch (Arca Swiss compatible) dovetail bottom. The wide rail alone promotes stability when resting on a surface, but add a shooting bag such as the Area 419 RailChanger that's easily positioned anywhere along its length, and you've got a complete rifle system in one hand that's perfectly stable on a fence rail, vehicle window, or the lowered ROPS bar on a tractor.

We don't want to wake the neighbors or spook livestock too badly, so we included Thunder Beast's titanium 223 Ultra 7 suppressor to keep noise and weight down. With a can and a light BCG, we were in for some tuning of the gas system. That turned out to be fairly painless using a Superlative Arms adjustable gas block. The SA block has positive click adjustments, making it easy to return the gas system to a prior setting after experimenting with a new load or rifle configuration.

The controls are fast and light to the touch. TriggerTech's new Diamond AR features a consistent 1.5-pound, two-stage pull that practically tells you when it's time to shoot. Adding to the speed of employment is Radian Weapon's 45-degree Talon safety.

GLASS

When choosing an optic, we wanted something light that'd take us at least to 10x magnification for 300-yard shots, but leaves enough at the wide end to scan the woodline effectively. We chose a Leupold VX-3i LRP 4.5-14 that runs in mils and has a first focal plane reticle so we can use holds at any magnification.

ALTERATIONS

For as well as this rifle handles and shoots, the build wasn't all lollypops and wet lips. We had to cut the anti-rotation tabs on the handguard to fit around the larger width of Grey Ghost's billet upper receiver. Instead of using the supplied set screws, we shaved the tabs' thickness so the bits that remain are in contact with the upper and prevent rotation.

We also had to hollow out the side of the Sierra Precision grip to accommodate the left-side safety lever. It turns out that the lever and grip work together as an excellent left-side thumb rest if you're into that grip style ... and we are.

PERFORMANCE

We tested a couple types of 22 Nosler factory ammo, Varmint 55-grain BT and Varmageddon 62-grain FBHP. The 55-grain load averaged 3,507 fps with a 0.49-inch five-shot group at 100 yards. The 62-grain load gave us 3,209 fps and a 0.46-inch five-shot group.

The whole build, can and glass included, weighs in at 10.75 pounds, with the base rifle and an empty mag weighing 8.4 pounds alone. The gun's just heavy enough to be a soft shooter with Proof's +2 gas length helping soften recoil a bit more.

The funny thing is this rifle really is a do-all, despite its length. The thing balances at the barrel nut and handles like a Porsche 911 with new tires. The hyper-fast 22 Nosler is devastating to animals and has legs that extend its use beyond 1,000 yards with less drop than 223 Remington and 224 Valkyrie. This guy could even be pressed into use as a PRS gas gun.

Whether from a fencepost or a pickup's bed, this 22 Nosler is quiet, accurate, and fast enough to deal with a whole pack of marauding coyotes, as long as we're as fast and accurate as it is. **R**

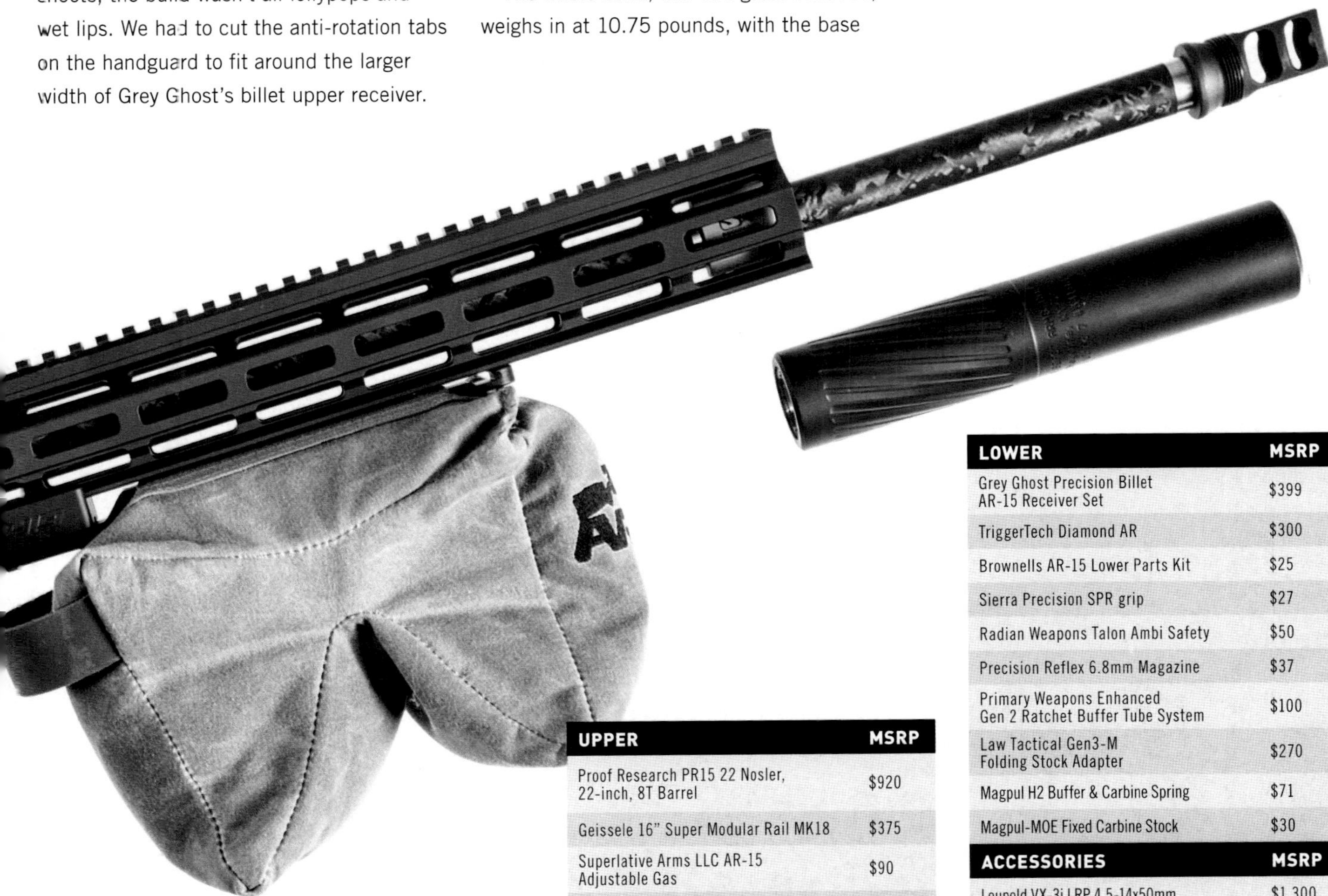

Using Law Tactical's folding stock adapter cut the admin length of the rifle substantially and allowed the benefit of running a fixed stock, the Magpul MOE Fixed Carbine stock.

UPPER	MSRP
Proof Research PR15 22 Nosler, 22-inch, 8T Barrel	$920
Geissele 16" Super Modular Rail MK18	$375
Superlative Arms LLC AR-15 Adjustable Gas	$90
Brownells AR-15 5.56 Titanium BCG	$260
Thunder Beast 223CB Brake	$125
Geissele Automatics Super Charging Handle	$89
V Seven Weapon Systems Ultra-Light Forward Assist	$28

LOWER	MSRP
Grey Ghost Precision Billet AR-15 Receiver Set	$399
TriggerTech Diamond AR	$300
Brownells AR-15 Lower Parts Kit	$25
Sierra Precision SPR grip	$27
Radian Weapons Talon Ambi Safety	$50
Precision Reflex 6.8mm Magazine	$37
Primary Weapons Enhanced Gen 2 Ratchet Buffer Tube System	$100
Law Tactical Gen3-M Folding Stock Adapter	$270
Magpul H2 Buffer & Carbine Spring	$71
Magpul-MOE Fixed Carbine Stock	$30
ACCESSORIES	**MSRP**
Leupold VX-3i LRP 4.5-14x50mm	$1,300
ZRODelta DLOC-M4 30mm Scope Mount	$279
Area 419 RailChanger	$175
Thunder Beast 223 Ultra 7	$995
Total:	**$5,958**

FIXED STOCK & NO SLOP

We Build Up from a KE Arms KP-15 Base

By Dave Merrill

The major standout of this build is the complete KP-15 SLT/Ambi polymer receiver from KE Arms. Featuring an integral A1-length buttstock, the KP-15 addresses a common weak point with other synthetic receivers — namely, breaking right where you attach your receiver extension tube and buttstock. While it may seem strange to rock a fixed stock in absence of a federal assault-weapon ban, the A1 length is fairly universal, and the taller among us can easily pop on a buttpad.

The KP-15 is also featured in the InRange/Brownells What Would Stoner Do (WWSD 2020) project. You may have seen a similar receiver from the now-defunct Calvary Arms, but the concept of an all-in-one polymer lower started with Eugene Stoner himself. While Stoner didn't find it viable at the time, we've had more than a few advances in materials science in the intervening decades.

Since this came as a complete lower, the internals are all KE Arms, with the excellent SLT-1 trigger included. The Sear Link Technology trigger maximizes energy transfer with the removal of the disconnector, allowing the selector to be placed on safe regardless of the hammer position, with a crisp break followed by a tactile, short reset.

The upper receiver is standard forged and came from a box o' parts. There's nothing special to see here, and the origins are unknown. The 4150 CMV, black-nitride, 16-inch, mid-length 5.56x45mm NATO 1/7 barrel was provided by DRG. Barrels are one of the categories where we sometimes have to scrimp and scrape during leaner times, so we were more than pleased to get our hands on a quality barrel.

The bolt carrier group is an AXTS (now known as Radian) black nitride model, with a hardened HPT/MPI 9310 steel

bolt. While original specs call for bolts machined from Carpenter 158, 9310 and S7 steels have proven to be just as durable in this application and are often easier to source.

While Griffin Armament is best known for their silencers, they have an entire line of AR small parts and furniture. For this build, we used the latest M-LOK Low-Pro RIGID rail — the longest they had available to maximize our real estate with a 16-inch barrel. We kept attachments pretty slick, only adding an Emissary Development M-LOK Handbrake. The Handbrake is larger than a finger-stop or index point but doesn't quite have the girth of a stubby VFG.

The silencer is a Gemtech Trek II. While it won't knock anyone's socks off with some new, yet unattainable level of sound reduction, the Trek II remains a relative short, solid performer with direct-thread attachment and corresponding wrench flats.

The optic mount is a Strike Industries Adjustable Scope Mount (ASM), and there are a few features we'd like to point out. The ASM can be set for one of four different eye reliefs to maximize rifle and scope compatibility, and ours features a preproduction ring top-half replacement, allowing for red-dot mounting. A universal MRDS mounting base can be attached at 90 or 45 degrees and can accommodate most footprints. And if you flip the mount over, you can attach an Aimpoint Micro.

Tiring a tad from the LPVO game, we upped our magnification game with the Lucid MLX. Featuring Japanese glass, mil/mil adjustment, and a first focal plane optical arrangement so the reticle scales along with the magnification, the MLX checks a lot of boxes at a lower price point. While 18x magnification is more than we'd usually ask for a 5.56 rifle, having that higher end can be useful for lower-probability shots or simply helping a friend check a target. A spare Atibal MRD V3 red dot was mounted forward and at 45 degrees for close-in shooting and as a secondary sighting method.

The only future additions we envision are perhaps a bipod for stability and a mounted LEP flashlight to maximize nighttime non-NV long-range shooting. It's not quite a kick-around do-all, but it's a well-balanced shooter biased toward the longer end of the spectrum. R

COMPONENT	MSRP
KE Arms KP-15 Complete SLT/Ambi lower	$450
White Label Armory/DRG barrel	$160
Upper receiver	~$105
AXTS/Radian bolt carrier group	$185
Griffin Armament RIGID rail	$210
Strike Industries ASM scope mount	$143
Strike Industries optical plate	TBD
Gemtech Trek-II silencer	$449
Atibal MRD V3 red dot	$289
Lucid Optics MLX 4.5-18x scope	$720
Emissary Development handbrake	$35
TOTAL	**$2,747**

DOWNSIZING

Sometimes Good Things Do Come in Small Packages

By Alexander Crown

We all know how it goes — you find that roll pin on the floor and now you have to build another gun. In this case, it was being gifted a really nice 9mm AR barrel that pushed everything over the edge to put this together. This build is something that had been pondering for considerable time. Now being "forced" to take action, some goals were identified: a compact 9mm, runs on Glock mags, and can be easily suppressed. This build is meant for fun ground squirrel hunting or occasional plinking when/if ammo prices even out.

COMPONENT	MSRP
Bootleg Inc Enhanced AR-15 Upper Receiver	$180
Spikes Tactical 9mm Spider Stripped Lower	$180
Bootleg Inc M-LOK 4-inch Handguard	$136
Primary Weapons Systems 9mm Bolt	Discontinued
Rainier Arms Select 6-inch 9mm Barrel	$150
Maxim Defense CQB Brace	$190
CMMG Lower Parks Kit	$69
Magpul MOE-K Grip	$19
Bravo Company USA Charging Handle	$40
EXTRAS	
SilencerCo Omega 9K	$865
Aimpoint H-1	Discontinued
LaRue Tactical LT660	$107
SureFire M600 Scout Light	$297
Magpul M-LOK Offset Light/Optic Mount	$35
KRISS MagEx2 Extension	$30
Glock 17 Magazine	$38
Edgar Sherman M81 Sling	$50
X2 Magpul Quick Detach Swivels	$30
Impact Weapon Components 2-to-1 TriGlide	$20
TOTAL	**$2,626**

The Rainier Arms Select Series Barrel was the perfect starting point for this little blaster. At 6 inches, it wasn't so long that subsonic rounds would go supersonic, since this build is intended to be shot suppressed almost exclusively. The barrel came threaded ½-28mm, making the mating of the SilencerCo Omega 9K easy with a direct thread adapter. The bolt was purchased at a steep discount locally due to the product being discontinued. It's from Primary Weapon Systems and features a removable carrier weight. The charging handle is the standard Bravo Company USA Gunfighter Medium, giving us increased surface area on the latch mechanism to help clear the sling QD swivel.

The Bootleg Enhanced Upper doesn't have a forward assist, which would be unnecessary for a pistol-caliber build, and it saves us a little bit of weight (1.7 ounces) over a standard Mil-spec upper receiver. The handguard is extremely easy to install, given the camming mechanism, and at 4 inches is a great length for the barrel. And, of course, M-LOK lets us mount stuff like the SureFire Scout light with Magpul mount to position the light higher at the 1 o'clock position for thumb activation. One gripe is the lack of M-LOK on the bottom, but we can get past this.

The lower receiver was purchased locally and is a Spikes Tactical "Spider" stripped lower. The lower included takedown pins and has a last round bolt hold-open device installed. A standard CMMG lower parts kit fills in the rest of the guts they may not be flashy, but the CMMG kit fit and didn't break the bank for quality parts. We'd like to upgrade the trigger to the CMC Single Stage 3.5lb 9mm PCC Flat trigger in the future. The Magpul MOE-K is an incredibly compact and thin grip that adds little in the way of weight and sits flush when a Magpul 21-round magazine or Elite Tactical Systems 22-round magazine is installed. An important consideration for storage of the gun.

To keep the overall length

of the build small and functional, a Maxim Defense CQB PDW brace was selected. These can come with a PCC weight buffer for use in blowback systems where more weight is necessary for reliable function. Once fully extended, the Maxim brace goes from 5.3 to 9.2 inches, providing sufficient length of pull for us for control.

The PDW brace also has quick detach mounts machined into the extension for attaching a sling. Since this package is small enough that a single-point sling would work, we selected the Edgar Sherman Design M81 sling with the use of an Impact Weapon Components 2-to-1 Point TriGlide. This little piece takes a two-point sling with quick detach mounts and allows you to make it into a single-point sling quickly. The sling also has the added benefit of an elastic band sewn in for storing the sling on top of itself for when the firearm is in a case or a bag. Topping off the build is the bombproof Aimpoint H-1 red dot riding in a LaRue Tactical quick detach, lower 1/3 cowitness mount.

The culmination of parts is a handy and capable package that fills the given parameters of being small and using Glock magazines. The benefit of Glock magazines is the ability to tailor the size to your needs, ranging from 10 and up to 50 rounds. This little guy tips the scales at 6.76 pounds, unloaded — not the lightest weight of builds, but this is acceptable given the heavy blowback components. Weight could be shaved replacing the PDW brace (but we'd sacrifice overall length) or using a lighter optic/mount combo. It's all give and take, and the important thing is that we have a fun package that can be upgraded as needed or wanted. **R**

6

Piston

You might assume by the nearby picture and title of this article that it's all about another AR variant. Like us, you're probably sick and tired of hearing about the "groundbreaking new" AR that, in essence, is the same as every other. Well fear not! Not only are you about to see firsthand what separates a company making dodgy gear from those companies making a quality product worth the extra bucks, but Adcor Defense's A-556 Elite operates completely differently from any AR rifle ever made. The Elite is the world's only free-floating gas piston AR — in fact, it's arguably the only truly free-floating AR on planet earth. You call bullshit? Well so did we. More on that later after an appropriate analogy.

When AR shopping, how are you supposed to tell the difference from so many similar-looking rifles? What makes the difference between a $600 entry-level AR and a quality, well-manufactured gat? Is that expensive rifle a quality product, or some well-marketed mediocrity by a company who just got themselves a CNC machine and barrel nut wrench, then slapped a graphic featuring a Spartan helmet and "Melon Labia" on the side of the receiver? And just what the hell is a melon labia, and would we appreciate one if we encountered it? All Latin jokes aside...

The necessary ingredients for a great

ADCOR DEFENSE A-556 ELITE

Hot, Filthy, and Suppressed, This Piston Gun Just Keeps Running

By Peter Palma
Photos by Shoot to Thrill

firearm lie in the company's ability to efficiently engineer products, precisely manufacture components, mass produce while maintaining quality control, and take feedback collected from the end-user and use it to improve their products. Adcor has the manufacturing boxes well and truly checked, because it's always been a precision manufacturer. The company specializes in making soda bottling machines, which run 24 hours a day, seven days a week, and are vastly more complicated than any semi-auto carbine. Adcor made the leap from soda-bottling machines to rifles when other major firearms companies approached them to manufacture AR internals faster and more accurately than they were capable of. Lest you doubt that soda-bottling equipment has any relevance to the business of making things that go bang, once word got around of its quality, orders for components for Trident nuclear missiles and F-16 radars started rolling in, so turning out a quality rifle is a comparative walk in the park.

During our tour of the facility, men in white lab coats were using large, complicated, and expensive equipment to measure tiny details of parts down to a gnat's ass. There was a room full of engineers quietly adjusting CAD drawings and poring over a myriad of numbers. Though engineers often look down from their

lofty perches on knuckle-dragging end-users like us, they're a necessary evil. They understand the black magic of physics and can calculate boring but critical things like loads and stresses that ultimately allow them to make rifles more accurate and reliable — and to use materials more efficiently, thereby cutting down on unnecessary weight, which, despite us being strong like tractor, we appreciate.

All of these factors mean nothing if the consumer is not a part of the process; otherwise you end up with a well-manufactured tool that's as useful as a baseball cap with sleeves. All too often, we see firearms or accessories with features that probably made perfect sense in the bright, clean environment of the design lab, but which fall on their ass when given to a cold, wet Marine who hasn't taken a shower in weeks and just ran out of dip. Oftentimes manufacturers, engineers, and QC technicians are not shooters, so it's important to include the end user in the feedback loop.

The A-556 Elite was one of five rifles selected from over 80 manufacturers to enter the final cut of the U.S. Army's Individual Carbine Competition before the powers that be canceled the program — a clear testament to whom this rifle is intended. Being end-users ourselves, we can see that the A-556 Elite is well thought out and well made. When disassembling and reassembling this rifle, the parts fit together perfectly, with pins and takedowns that are cleverly captured and require no tools to operate, besides a 5.56mm round.

PASSING GAS FREELY

Though the A-556 Elite looks like a run-of-the-mill M-4, what exactly separates it from the herd? Just the fact that it's piston driven and free floating! You may want to sit down while we explain this. Like any other gas-operated rifle, there's a hole in the barrel through which gasses are redirected to cycle the action. In all other gas-piston ARs, these gasses push a piston, which then transfers rearward force to the bolt carrier. The same happens in the A-556 Elite, but in this case there's a ring around the barrel with a probe that diverts gas into a separate chamber, where the short-stroke piston is held in place by the proprietary handguard system. This piston then pushes rearward on a one-piece, proprietary bolt carrier.

At no point does the barrel touch any part of the rifle other than at

the barrel nut, and the entirety of the piston system is held in place by the handguards mounted solidly to the upper receiver. The gap between the gas probe and the short-stroke piston chamber is like that between a revolver cylinder and barrel — no contact is made. Even in so-called free-floating direct-impingement ARs, the barrel isn't truly isolated because the gas tube is still connected to the gas block and threads back into the upper receiver. The A-556 Elite doesn't even have a gas block. This system also allows the bolt carrier to have multiple points of contact in the upper, completely eliminating carrier tilt, common among other piston-operated ARs.

So does this new system work? Why yes, yes it does.

We put a target at 100 yards, dropped down into the unsupported prone, peered through our three-power ACOG, and consistently achieved 1-inch groups with both Hornady TAP and 69-grain Federal remanufactured ammunition. Even with a 14.5-inch cold hammer-forged chrome-lined M4-profile barrel using less than match-quality M-855 ammo, we still easily produced groups of less than 1.5 inches. After being bored stupid with its consistent accuracy, well, you know what comes next. Yep, we shifted our focus toward trying to break the thing.

Despite throwing it around and covering it in mud (especially around

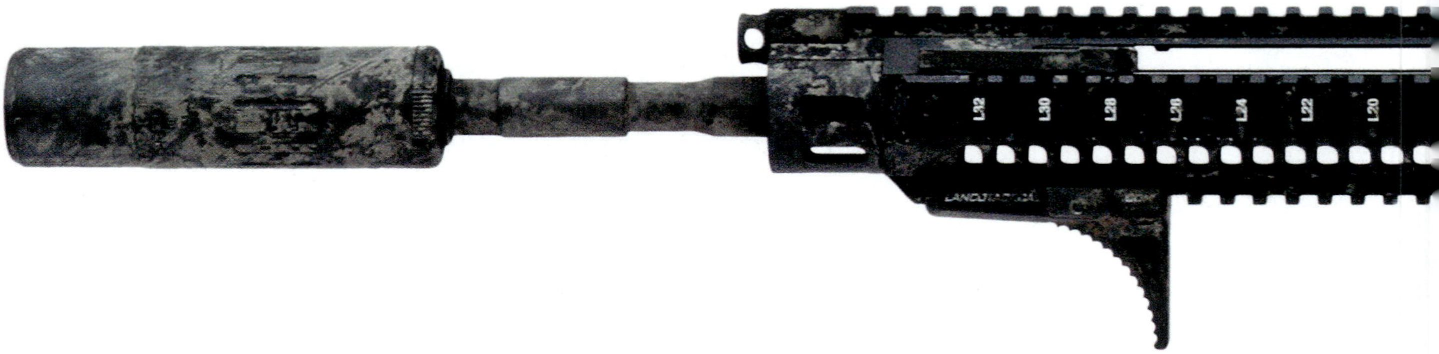

the open area of the handguard) there were no malfunctions. We even went against our better judgment and filled the barrel with swamp water (as seen on the next page) and fired it. There were zero malfunctions with any of the multitude of ammo types we used, making it one of the most accurate and reliable M4s we've tested. However, the Elite's reliability didn't really come as that much of of a surprise to us. When touring the factory's test range we noticed a pile of filthy rifles and drums filled with various materials such as dirt, sand, mud, water, and hobo feces. OK, we lied about the last one. Staffers encouraged us to get the rifles as filthy we wanted and shoot the snot out of them to try to induce a malfunction. Short of opening the action and filling it up completely, we were miserably unsuccessful in our attempts to stop the A-556 Elite from going bang.

FORWARD THINKING

The unique gas system is also adjustable for suppressor use, different ammunition types, or to change the cyclic rate. When fitted with an Advanced Armament Mini-4 set on the lowest of the four gas settings, our test rifle ran clean and cycled flawlessly. We also had no excess gas come back in our face. Stop laughing.

The short-stroke piston, where the vast majority of the fouling occurs, can quickly and easily be removed for cleaning, thus keeping the rifle running longer with minimal amount of effort. The A-556 Elite also has a non-reciprocating forward charging handle cleverly tucked away between the top and side rails so it won't snag. However, it's easily accessible and automatically folds forward under spring tension. For the sake of argument, let's say you could afford the ammo budget to completely cake the piston with carbon and turn the rifle into a single shot. You could still effectively engage multiple targets because the ergonomic position of

ADCOR DEFENSE A-556 ELITE

CALIBER: 5.56 NATO

MAGAZINE CAPACITY: 30 rounds

BARREL LENGTH: 14.5 inches, plus pinned and welded muzzle device

OVERALL LENGTH: 31 inches (stock collapsed)

WEIGHT: 6.85 pounds

MSRP: $2,200

URL: www.sionicsweaponsystems.com

FEATURED ACCESSORIES:
B5 Systems Grip Stop - $55
Trijicon 3x30 ACOG - $1,407
AAC MINI4 suppressor - $896

PRICE AS FEATURED: $4,558

the forward charging handle allows you to maintain sight alignment while manually cycling the rifle.

Not only is the forward charging handle ambidextrous, but unlike other similarly equipped carbines, it complements the standard rear charging handle that's located in the same place it's always been since Stoner finalized the design. If, like us, you've built up muscle memory for the M4's manual of arms, then there's nothing to stop you from using the gun in the manner you're used to. On the downside, the forward charging handle sits so close to the Picatinny rail that nothing can be mounted on the same side, though it can be removed all together without tools, should the shooter need that particular bit of real estate.

Not only does the ejection port have the standard dust cover, but the bolt carrier has another interesting feature to help prevent debris from entering the rifle. A radiused piece of self-lubricating polymer is attached to the bolt carrier via two springs that push toward the ejection port. This dust swipe closes all gaps around the ejection port leading into the rifle, even if the cover is not closed.

The A-556 Elite is available in a number of barrel lengths. The carbine-length handguard version comes with either a 16- or 14.5-inch barrel, for those who choose to permanently attach a muzzle device to bring the overall length of the barrel up to the legal minimum. If you choose to go the SBR route, a 10.5-inch barrel is also available. For those who use the Costa Lotus-style grip with your support hand, a longer rifle-length rail is available in 18- or 20-inch barrel lengths. All A-556 Elites are optic-ready with no fixed front sight and are chambered in 5.56mm. All upper receivers, lower receivers, and handguards are machined from billet 7075-T6 aluminum.

The Adcor A-556 Elite is the product of a perfect storm of state-of-the-art manufacturing, great engineering, attention to detail, and forward thinking. For every job, there's an ideal tool. The A-556 Elite doesn't have a cool name or even look particularly special. It's black and doesn't come in the latest popular color or camouflage pattern. It doesn't have the new handguard with a new accessory locking system, or a fancy super-loud muzzle brake, or even amusing pictograms on the selector lever and ejection port cover. It isn't meant to sit unscratched in the gun safe until it's time to impress your friends, it's simply a tool. It's for those shooters who require a superbly accurate M4 designed to run with limited maintenance, under the most adverse conditions, even while suppressed. What's not to like? R

PISTON

We Get Intimate With Faxon Firearms' Switch-Barrel AR-15 Upper Prototype

By Peter Palma
Photos by Shoot to Thrill

What does the Mars Rover have in common with a piston-driven AR? Read on and find out...

Despite their recent surge in popularity, piston-driven ARs have been around for a long time. Colt was the first company to try solving the drawbacks of the original direct impingement (DI) operating system, way back in the decade that good taste forgot. Since the '70s, numerous others have jumped on the bandwagon, with most of them riffing on Colt's theme of a short-stroke piston acting on a modified AR bolt carrier. Due to design constraints imposed by the DI upper receiver, all of them swap the old problems of heat and carbon buildup for the new issue of carrier tilt, with some manufacturers addressing it more effectively than others.

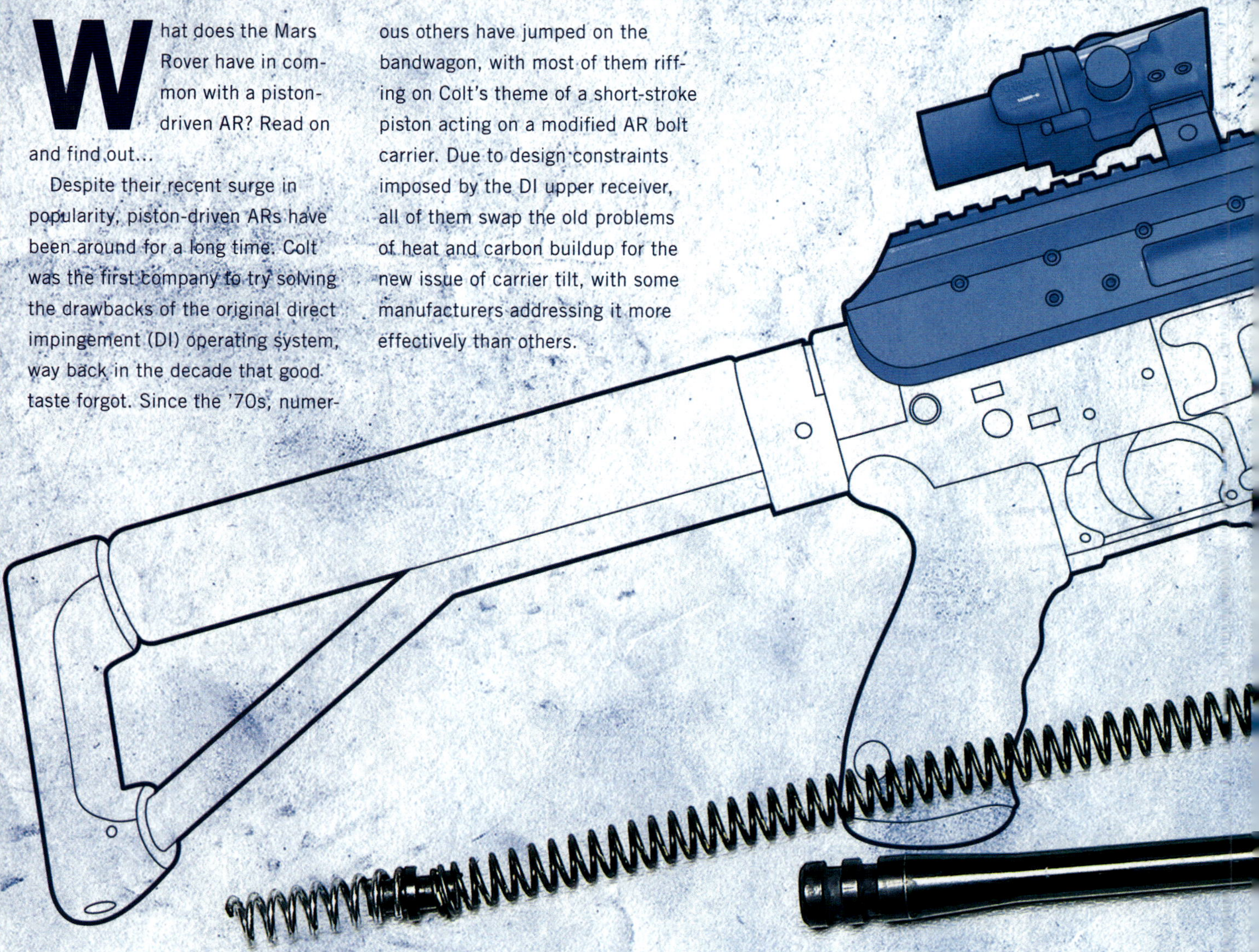

POWER

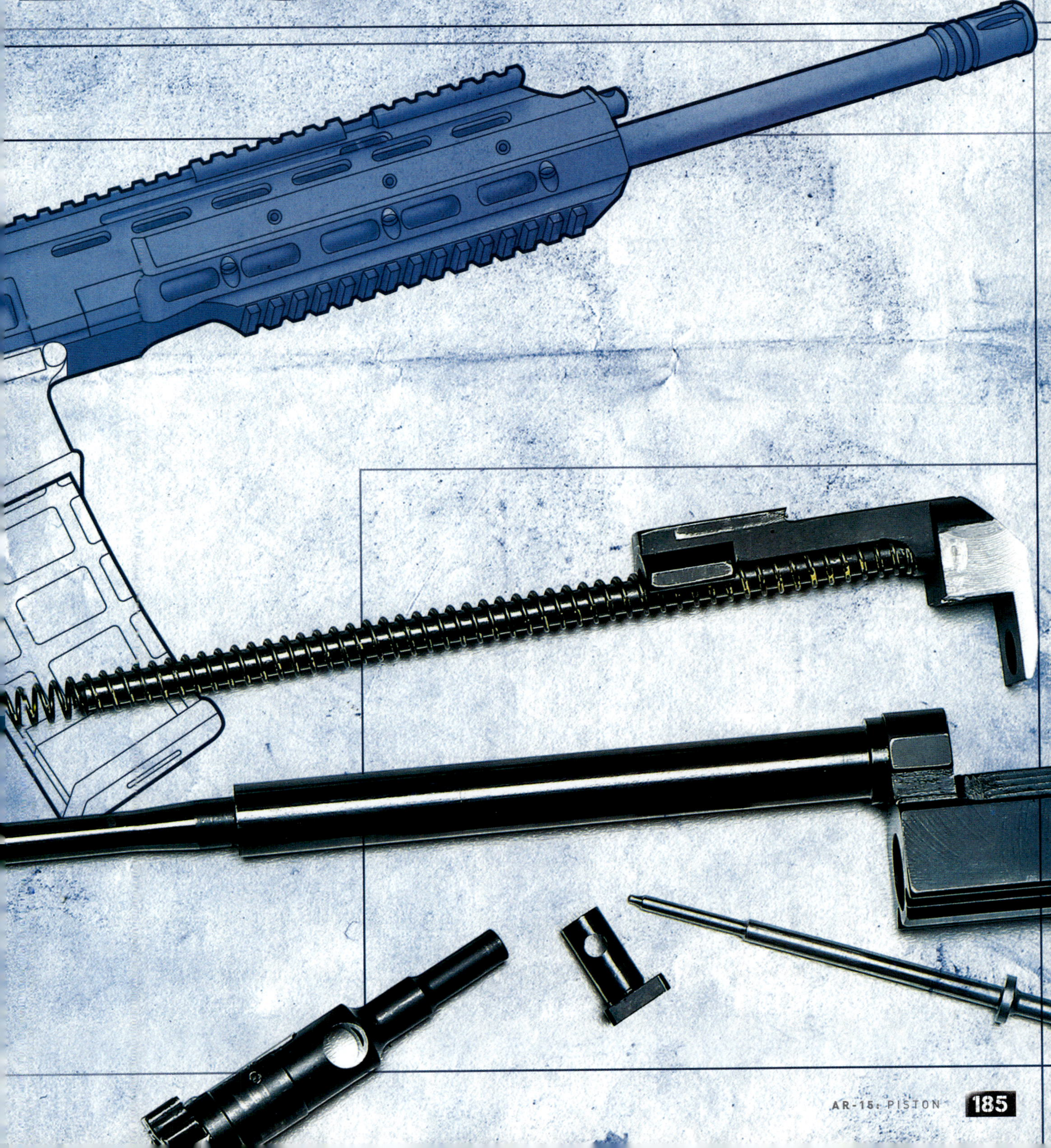

Solving hard engineering problems is what Bob Faxon does for a living. Faxon's company developed the parachute system for the recent Mars Rover, so you could say he has some experience in building gear that works when there's no chance for a do-over. Bob's also a shooter and it was only natural for him to found Faxon Firearms. When he looked at existing piston AR designs, he figured there had to be a better mousetrap. The biggest problem any AR piston gun has is caused by trying to cram a different operating system into the original upper receiver, so this was the first thing to be ditched. Once freed from its constraints, Faxon was able to design an upper that incorporates hardened, full-length rails for the carrier to ride on, completely eliminating the problem of carrier tilt. The new receiver design also had enough room inside it to accommodate a recoil spring, removing the need for the AR's buffer tube and making a folding stock possible.

We tested the first preproduction prototype to be let out of the Faxon shop, and there's a lot of innovation and hard work gone into this design, so let's examine it from front to back.

The muzzle device is an A2 flash hider with standard threads, so a screw-on suppressor will fit without modification. Two barrel profiles are due to be offered: a fairly hefty medium contour and a lightweight pencil barrel, both in 16-, 18- and 20-inch lengths. The 5.56-caliber barrels will feature a 1:7 twist and the 300 BLK will utilize a 1:8, so that the widest range of bullet weights can be used and so the user can practice with lower-cost ammo before, say, heading out to slay hogs with the more expensive chambering. Both barrels are made in-house from 4145 pre-hardened chrome-moly steel and there are plans to nitride the production versions (along with the bolt and carrier) to increase their life span.

An adjustable gas block is permanently installed 9 inches from the chamber and has settings for normal and suppressed operation. Because this is a piston gun, the benefits of a free-floated barrel can't be fully realized, so the Faxon team took the approach of keying the gas block into slots cut into the monolithic upper. The theory is that this approach stabilizes everything to the rear of the gas block and adds rigidity to both the barrel and handguards, but because the key slots allow for independent expansion as everything heats up, point-of-impact shift is minimized.

We did notice some vertical stringing on the range when using 77-grain bullets in the 5.56 upper, but 55-grain bullets all shot into a circular group. Accuracy was about average for a piston gun and with this sample we could count on five-shot, 2-inch groups at 100 yards all day long with ammo that it liked, with some cloverleaves thrown in for good measure.

TAKE CHARGE

Riding in a slot cut in either side of the handguard is a forward-mounted charging handle, which can be switched for either right- or left-handed use. Unlike the one used by the SCAR elsewhere in this issue, the Faxon's is non-reciprocating. This is a big improvement in ergonomics over the AR's charging handle location, which was barely OK when originally designed in the iron-sight era, but (let's face it) sucks when using optics. In the event of a stoppage, the user barely has to shift the support hand grip in order to rack the charging handle and get the gun back into action. The upper receiver features a one-piece top rail for optics mounting (rail height is approximately ½ inch taller than a typical AR). The barrel is retained by a lower handguard that attaches via six captive screws, which thread into steel inserts rather than relying on just the aluminum upper for strength.

A long-known weakness of the AR platform is its bolt, which has sharp corners at the base of each of lug, producing stress risers that eventually cause the lugs to shear. One way

411 ON PISTON AR'S CARRIER TILT

So what is this carrier tilt thing and why does it matter? To answer this, we should look at the beauty of the direct impingement (DI) system and how it operates. When you drop the hammer on a regular AR, the primer lights off the main charge of propellant, which generates a large amount of high-pressure gas. This gas pushes the bullet down the bore on its way to the target, expanding as it goes. When the bullet passes the gas port, some of this gas makes a swift right turn and enters the gas tube, passing along it and through the gas key on the bolt carrier, into the space between the bolt tail and carrier body.

The expanding gas then forces the bolt and carrier apart — the bolt can't go forward, but the carrier can move rearward and so it does, starting the cycle of extraction and feeding. Note that the gas key only acts as an extension of the gas tube; it isn't a piston. All the action happens inside the bolt carrier group and everything acts in a straight line.

Now stay with me here, as we're almost done with the dull engineering lecture. Piston ARs no longer have everything acting in a straight line. In most designs, the piston system rides above the barrel and smacks into an extension on the top side of the bolt carrier — and when it does, the carrier gets driven not only backwards, but also downward. If you take a gander at your bolt carrier, you'll see two short bearing surfaces on each side of the gas key and a couple of matching ones on the bottom of the carrier. That's all the BCG rides on inside the upper receiver, which is fine when all the forces are acting in a straight line, but it leaves the end of the carrier flopping around if it gets whacked by a piston. From an engineering standpoint then, this state of affairs is less than ideal.

Above: Faxon Firearm's ARAK 21 upper completely stripped. Note the AK-style carrier and AR-style bolt.

Left: Folding, forward-mounted charging handle. The upper receiver has mounting points for additional rails at 3 and 9 o'clock positions.

to prevent this is by turning those sharp corners into a gentler radius, spreading the forces out instead of concentrating them. Rather than using a typical AR bolt, the ARAK's bolt and the matching barrel extension feature lugs that are radiused over their entire length, producing a design that the manufacturer claims is 30 percent stronger. The ARAK bolt also increases the number of lugs from seven to eight, though the pair that straddles the extractor is shaved down a little. The bolt lugs most susceptible to shearing in the AR-15 platform are the ones closest to the extractor, as they take much more stress than the others due to asymmetrical spacing around the bolt head. The ARAK's bolt lugs are more evenly spaced, as well as being closer together, and hence, more equally stressed.

About the only criticism that can be leveled at this arrangement is that the extractor's bearing surface is reduced versus that of a normal AR. Having ripped extractors through a fair number of case rims in our time, we reserve judgment on this design until we've been able to put several thousand rounds through the upper. (Note: we were recently shown samples of a revised bolt head that featured a wider extractor, but did not have one on hand for this test).

CARRIED AWAY

The heart of the Faxon system is its carrier, which is a work of industrial art. Comprised of two main sub-assemblies, namely the piston and the carrier itself, it serves as a sleeve for the recoil spring, much like an AK-47, while housing the very AR-like bolt. Disassembly couldn't be easier. Press down on the lever between the guide rails, tip out the standard AR firing pin, lift the cam pin and pull the bolt out of the front. There's no cotter pin to lose. Faxon plans on offering the upper in both a right- and left-handed version and, because of this, it's possible to reassemble the bolt with the head

The upper receiver is hogged out of a solid block of aluminum. Lower handguards secure barrel and gas system to upper.

The Faxon upper comes with both 5.56 and 300 BLK barrels and fits on any in-spec AR-15 lower.

oriented 180 degrees out of position. I did not do this, nosiree...

The system's recoil spring is well thought out. With the outside diameter enclosed by the piston and a guide rod running up the center, there's no possibility of kinking and service life is probably measured in decades. The last inch of spring is made from a heavier gauge wire, which cushions the carrier's final few thousandths of travel and reduces felt recoil. The whole system bottoms out on the guide rod, which means the spring always has air between its coils and is never overstressed.

PERFORMANCE EVALUATION

We got the opportunity to run the gun hard (not just put a few rounds through it at a press event), and as is the case with all prototypes, there are a few things that we would change. Firstly, it's possible to tie the gun up if you press in the takedown plate in the rear of the upper receiver when everything is assembled. When this happens, the user has to pivot the receivers apart and reset the plate back to its working position, which is decidedly sub-optimal. We've learned that a modification will be added to the plate to prevent this from happening in production uppers, so this should hopefully be a non-issue.

Markings for the gas settings should be engraved on both the gas block and plug, and these should preferably be tactile also. The biggest issue from a usability standpoint, though, relates to the monolithic upper and its relationship to the gas block. By tying the two together, heat is transferred to the handguards. After dumping a Surefire 60-round mag, this area gets hot. Like, swearingly hot. By adding side rails and KAC covers, or by running the gun using a VFG, this can be overcome, but for now it's a

Left: Controllable and fast-cycling, the upper balances well with weight concentrated between the hands.

problem. Finally, our prototype was tattooed on its left flank with only slightly fewer graphics than a NASCAR starting grid. Not a fan.

Overall, we think that this is a very promising design and would like to see it in a couple of other options. The 6.5 Grendel has always run on the very edge of the AR's performance envelope and the Faxon bolt seems to offer a worthwhile upgrade in terms of strength. The lack of a buffer tube would lend itself to a very slick, folding stocked, 10-inch barrel PDW variant, particularly in 6.8 SPC, as that cartridge really shines in short barrels.

By offering a high-quality piston upper with both 5.56 and 300BLK barrels for $1,459, we predict these will sell like ice cream on a summer day, especially among suppressor owners or those looking for a bit more downrange performance than the NATO cartridge offers, and we look forward to seeing how the design matures. For now, we'll be holding onto this one and trying our best to break it over the next few months. Check back for updates. R

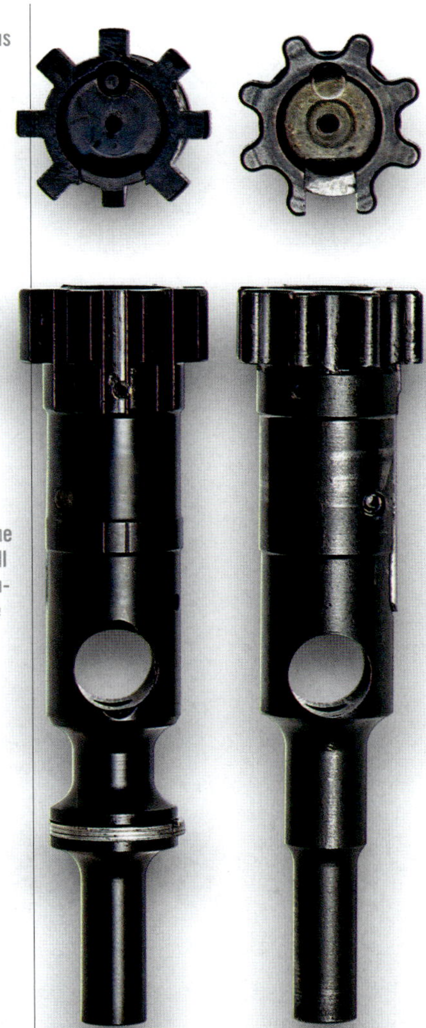

Right: AR bolt versus Faxon bolt. The wave-form lugs should be stronger and more durable.

FAXON FIREARMS ARAK 21

CALIBER: 5.56 NATO, 300 BLK

MAGAZINE CAPACITY: n/a

BARREL LENGTH: 16, 18, and 20 inches

OVERALL LENGTH: 24.5 inches (with a 16-inch barrel)

WEIGHT: n/a

MSRP: $1,459

URL: www.faxonfirearms.com

ADAPTABILITY:
CONVERSION TO PDW

To make the most of the Faxon's features and to address its shortcomings, we kitted it out with a folding stock and VFG, which served to reduce the carbine's OAL to 26.5 inches and negate the problem of the handguards heating up. The vert grip came from CTC Defense and while spendy, offers the user the ability to swap out laser and illumination modules for both visible and IR targeting, as well as slaving other electronics from the unit. Trijicon's SRS red dot, described elsewhere in this issue, completed the optics package and also offers NV capabilities to compliment the IR laser.

The folding stock lacks a useable sling mount when folded, so in order to add a single point sling, we resorted to the tried and true method of a paracord loop, which worked just fine. Hey, at least we didn't need any duct tape...

INCLUDED ACCESSORIES:
Upper Receiver: Faxon Firearms ARAK-21
Lower Receiver: Del-Ton Inc. DTI Stripped Lower Receiver
Foregrip: Crimson Trace Corp. MVF-600C
Buttstock: ACE Ltd. Ultra Light Modular Fixed Stock
Side-Folding Adapter: US Machinegun Side Folding AR/ Tube Stock Adapter
Pistol Grip: Ergo Grip Suregrip (ambidextrous)
Rail Cover: Ergo Grip 18-Slot Lowpro Ladder Rail Cover
Optic: Trijicon SRS
Magazine: Magpul PMAG 30 AR/M4 GEN M3 Window Magazine 30-Round

FAXON FIREARMS OUR PERSONAL DEFENSE WEAPON CONVERSION

CALIBER: 5.56 NATO, 300 BLK

MAGAZINE CAPACITY: 30 rounds

BARREL LENGTH: 16 inches

OVERALL LENGTH: 26.5 inches

WEIGHT: n/a

PRICE WITH BOTH BARRELS AND ACCESSORIES: $5,033

URL: www.faxonfirearms.com

Ammo was fed from the latest generation of Magpul's signature product, which has been redesigned to give increased compatibility with a range of non-Colt spec magwells. The Gen 3 Pmag now includes an overtravel stop, so that when you slam it in on an open bolt, there's no chance of jamming it past the mag catch and tying up the gun. The floorplate is also a bit slimmer, which makes pulling a reload out of your mag pouch a little easier.

Now, imagine the Faxon with the barrel cut down to 10 inches. ■

Warning:
Read
operators
manual
before
use
Heckler & Koch Inc.
Columbus GA

HK MR556A1

Civilian Version of the Professional's Choice

By Dennis Ideue and Recoil Staff
Photos by Recoil Staff

By all credible accounts, the Heckler & Koch HK416 was the weapon selected by DEVGRU (SEAL Team Six) to avenge the United States in the daring raid on Osama Bin Laden's hideout. In addition, Delta Force selected the HK416 to replace its M4s. Those are pretty astonishing endorsements, so we were excited to get our hands on the civilian version: the HK MR556A1.

Heckler & Koch (HK) has always been fanatical about precision machining, accuracy and repeatability, which are almost expected from a German manufacturer. As with the German transportation system, HK rifles operate like a precision clockwork mechanism.

The MR556A1 is HK's first commercially available AR-platform rifle. It uses a short-stroke, gas-piston operating system that is derived from the company's successful G36 rifle. With piston systems, hot gases do not enter the chamber to cycle the firearm, which keeps the chamber and the components housed inside cleaner and cooler. This system has been shown through testing to provide easier maintenance and increased service life of internals and increased time between service intervals. Those who obsessively clean every weapon after range sessions always welcome gas-piston designs.

1 The rounded edges of the lugs help increase reliability.

2 MR556A1's proprietary HK bolt carrier.

3 An HK-marked buffer rounds out the operating system.

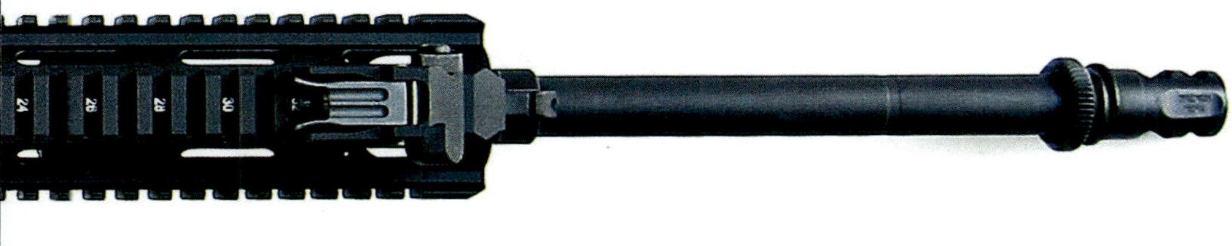

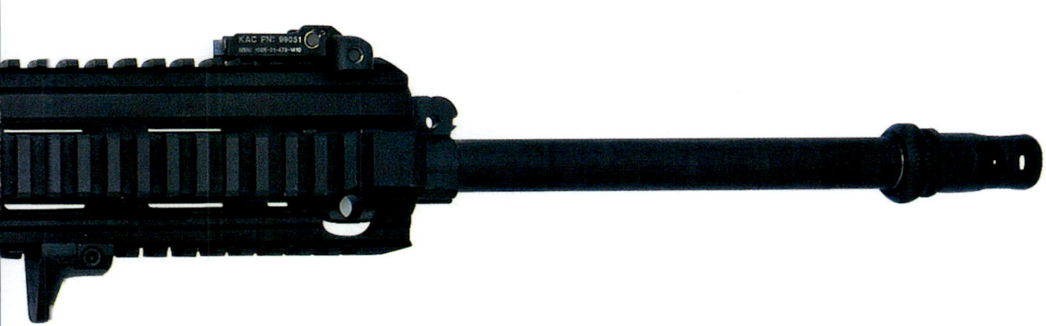

HK's proprietary short-stroke system includes the operating rod and piston situated over the barrel, where a traditional AR's gas tube would be. The rod actuates a rotating seven-lug bolt, with rounded lugs housed in a special HK bolt carrier. The rounded edges of the lugs help increase reliability. A proprietary action spring and HK-marked buffer complete the system.

The MR556A1's barrel is cold hammer forged and 16½ inches long, with a heavy profile. The twist rate is 1:7, and there is no chrome lining. HK engineers choose to not chrome line the barrel because they believed it could mask imperfections that could reduce accuracy. The barrel has, as HK describes it, a "swagged" muzzle end. This means that the internal diameter of the barrel at the muzzle is slightly smaller than the diameter at the chamber end. HK cites this "swag" as giving the barrel better projectile accuracy and velocity.

GERMAN DESIGNS CAN BE OVERLY COMPLICATED.

Surrounding the barrel is a 9-inch-long full free-floating quad Picatinny rail. To accommodate the rod and piston, HK raised the entire rail and upper receiver almost ¼ inch. This can affect optic mounting choices. Larue Tactical is one of a couple of companies that makes specific MR556/HK416-height optic mounts.

As with many German-engineered products, the machining work on this rifle is exemplary; each component seemed to be machined with loving care. The MR556 shares components made in Germany and the United States, and final assembly is done in the U.S. The fit and finish of this rifle are top notch, but there are other features that we did not like.

German designs can be overly complicated. Take the MR's front and rear takedown pins, for instance. They are captive, which we found to be a real pain and quite unnecessary. A special tool, which has a specific storage area in the butt stock, is used to depress a detent in the center of the pins in order to unlock them before they can be pushed out or back in. We see no good reason

4 MR556A1's internally "swagged" barrel gives more projectile accuracy and velocity.

5 To accommodate the gas piston, the rail and upper receiver are raised almost ¼ inch.

6 Companies such as LaRue Tactical make shorter HK-specific optic mounts. We intentionally used a raised mount to help clear an IR laser that can be seen on the cover of this issue.

7 A special tool is needed to unlock the takedown pins.

8 The takedown tool is stored in the butt stock.

1 The extended latch on the charging handle.

2 The ejection port cover is made of plastic.

3 HK's E1 butt stock can be found on the MR556A1 and HK416.

4 The E1's length-adjustment latch can be awkward to operate.

5 There's plenty of storage space inside the butt stock.

6 The proprietary HK grip has a steeper angle than traditional AR grips.

7 Magpul's EMAG fit perfectly in the MR556A1's tight magazine well.

8 Unlike other AR-platform rifles, this gun can be put on safe even without the hammer cocked.

why another two steps and an extra tool were added to the breakdown and reassembly of the rifle. And if you want to swap to an aftermarket stock, you'll need to find a new place to stow your takedown tool. Luckily, standard milspec takedown pins can be used and would be one of the first modifications we would perform.

HK includes an ambidextrous safety-selector switch, which is a nice touch regardless of whether you are right- or lefthanded. The extended latch on the charging handle is definitely a plus for easier and more efficient clearing manipulations, as well. But the ejection port cover is unexplainably made of plastic rather than the steel that is found on the HK416 and all other ARs.

The MR's HK E1 stock is found on the military version HK416 as well. It is collapsible and is fitted over a six-position milspec-size buffer tube. The stock has a generous storage cavity that can hold up to six 123-size batteries as well as the takedown tool. One gripe we have with it is that the adjustment lever can be troublesome to use. The sides of the lever are framed by the stock so that adjusting it means that you need to press the lever down with your fingers situated in a particular way. Those who have large hands may find this particularly bothersome. The stock does include a screw that can lock it into a fixed position for a more solid feel, and it also includes a comfortable recoil-absorbing rubber butt pad.

The MR also shares grips with the HK416. The grip is said to be ergonomically shaped, but our staff didn't find it so. Fans of personal defense weapon (PDW) style grips might enjoy the steeper grip angle that the HK grip has, but we found that the angle on the AR-platform gun just didn't feel right to us. A lot of users have retrofitted the MR with aftermarket grips because of comments about the HK grip feeling "different."

Since the MR556 is of the AR-family architecture, most of the AR add-ons fit perfectly. Some of our favorites worked flawlessly, including the Magpul Battery Assist Device (BAD) lever.

HK supplies its legendary steel-construction High Reliability magazines with the MR. If you've ever held one, you understand just how amazingly strong they are. But with that strength comes heft. The high-reliability magazine weighs more than double what a regular (albeit less durable) aluminum USGI magazine weighs. We chose a happy medium to outfit our MR with—the sturdy and lightweight Magpul EMAG. The more common Magpul PMAG will not fit the MR's tight magazine well.

One feature that HK nailed but no one else to our knowledge has done with the AR platform is allowing the safety selector to be switchable to the

WE CHOSE A HAPPY MEDIUM TO OUTFIT OUR MR WITH—THE STURDY AND LIGHTWEIGHT MAGPUL EMAG.

safe position even without having the rifle charged. Typical ARs cannot be placed on safe unless the hammer is cocked. As simple as it sounds, this safety upgrade over conventional AR platforms is something we really like.

To test the MR, we set up a course of fire that involved run-and-gun scenarios with paper targets 5 to 25 yards away and steel gongs at 35 and 50 yards. We outfitted the MR with an Aimpoint Micro T-1, AAC Blackout muzzle brake, Knight's Armament Front and Rear Flip-up sights and Gear Sector Hand Stop.

The feel of the gun is much like other gas-piston-operated carbines. A sharp impulse hits you with every trigger press, unlike with non-piston ARs. This impulse made the dot jump, and it took a split second to get back on target.

The more we ran the course, the more apparent the weight of the rifle became. An out-of-the-box MR556A1 weighs in at more than 8 pounds, and our setup with accessories tips in at 9 pounds—unloaded. Add another pound for a loaded magazine, and we're talking an operational weight of 10 pounds. That's like running around with a bowling ball held high against your cheek. It could get pretty brutal over extended periods.

In the role of a close-quarters combat type of carbine, which the military and law-enforcement-only HK416 is intended to be, the civilian MR556A1 is on the heavy side. Most of its weight lies in its heavy-profile barrel.

There are positives for the MR's heavy-profile barrel, though. When the rifle is being shot out to longer distances, it performs like a match-grade weapon. Though not quite equivalent to a 20-incher, the 16.5-inch barrel let us produce consistently tight groups using Remington UMC 55gr ammunition.

As mentioned earlier, the recoil was more abrupt than with other AR platforms or even the gas-piston FNH SCAR 16. When supported on the bench or prone on the ground, the rifle's weight helped stabilize the gun for more precise shots with minimal muzzle jump—again a byproduct of the heavier barrel.

Build quality and cost place this rifle in the upper end of the AR variants. Pricing may vary from reseller to reseller, but you could feasibly assemble two nice ARs for the cost of one MR556. Like a Rolex watch or an ST Dupont lighter, this is not every man's rifle. It was built and designed to be one of the most reliable AR platforms in existence, and reliability—especially German reliability—comes at a price.

Overall, MR owners seem to be thrilled with their purchases. Gun forums are filled with glowing reviews and comparisons to tanks and bank vaults when commenting on the fit and finish of the components. Most RECOIL staffers enjoyed shooting the MR556A1, and all of us enjoy looking at it. While it is certainly a beautiful rifle, the ergonomics and quirks of it had many of us wishing HK went with more traditional controls.

MR556A1 pricing is higher than with other gas-piston AR-style rifles such as the PWS MK116, Ruger SR-556 and the LWRC M6 series. But with the MR, you are getting one of the best-crafted AR-platform rifles that we have ever encountered. **R**

HECKLER & KOCH MR556A1

CALIBER: 5.56 NATO
MAGAZINE CAPACITY: 20 or 30 rounds
BARREL LENGTH: 16.5 inches
OVERALL LENGTH: 33 11/16 inches, stock collapsed
WEIGHT: 8 pounds, 5 ounces
MSRP: $3,295
URL: www.hk-usa.com

FEATURED ACCESSORIES:
Aimpoint Micro T-1 4MOA - $633
LaRue LT660 QD Mount - $98
AAC Blackout 51T Muzzle Brake - $150
Gear Sector Rail-Mount Hand Stop - $48
Gear Sector Rail-Mount QD Swivel - $48
Knights Armament M4 Flip-Up Front Sight - $190
Knights Armament Micro Flip-Up Rear Sight, 200-600m - $153

PRICE AS FEATURED: $4,615

HK MR556A1 vs. HK416

DIFFERENCES BETWEEN RESTRICTED AND CIVILIAN HK 5.56MM RIFLES

There's no doubt that there is major "cool factor" in owning a cutting-edge, military-issue firearm, let alone one that reportedly took down America's most wanted terrorist. The Heckler & Koch HK416 was developed in conjunction with members of U.S. special-operations units and is currently being used on the front lines of the Global War on Terror. The closest a civilian can get to owning an HK416 is Heckler & Koch's MR556A1. How close does the MR556A1 get to the military and law-enforcement-only version HK416? Let's take a closer look.

The bulk of the parts that make up the HK416 and MR556A1 are the same, and both run HK's proprietary gas-piston operating systems. There are a few key differences, though, most notably in the lower receiver. The HK416's fire control group is capable of full-auto operation, whereas the MR556A1's two-stage trigger is capable of only semi-auto fire.

Aside from being denied the fun of full-auto fire, the civilian version MR556A1 is a very close match to the original HK416. It uses the same proprietary free-float quad rail, and both use components that are made in Germany. Unlike the HK416, the MR556A1 also uses some American-made parts and is assembled in the U.S. Even so, craftsmanship on both models exudes German precision, and both are clear works of Teutonic art. If you want an HK416 of your own, you will definitely want to take a good hard look at the MR556A1. **R**

1 Both the HK416 and MR556A1 have German-made cold-hammer-forged barrels, but the HK416 (bottom) we studied had a government-profile, 14.5-inch chrome-lined barrel. The MR556A1 (top) comes with a 16.5-inch non-chrome-lined, heavy-contour barrel. Also, the bayonet lug has been taken off the gas block on the civilian MR556A1.

2 There is a protruding lug in the MR556A1's barrel extension that makes it impossible for a full-auto HK416 bolt carrier to be used.

3 A negative space is cut into the semi-auto bolt carrier of the MR556A1 to accommodate the barrel extension's protrusion. The HK416's bolt carrier does not have this cut.

4 Locking takedown pins hold the upper to the lower on the MR556A1. The HK416 uses standard AR-platform takedown pins.

5 The ejection port cover on the HK416 is metal, while the MR556A1's is plastic.

1 MINUTE PER REVOLUTION

POF's Newest Piston Gun Delivers AR-10 Performance in an AR-15 Package

Story by Iain Harrison
Photos by Kenda Lenseigne

With every man and his dog making ARs these days, it takes something rather unusual to make it into the pages of RECOIL. Patriot Ordnance Factory has been teasing the 7.62 NATO Revolution for some time now, and we were given a sneak peek over a year ago while it was in its direct impingement (DI) form. Now that a piston-driven production version has been released, we decided to see if it lived up to the hype.

The Revolution's main claim to fame is that it manages to cram 7.62 performance into a 5.56 form factor. Separate the receivers, and you'll find that the Revolution upper will drop right onto a Mil-spec mousegun lower. The reverse isn't true, however, as extra material in the 7.62 lower's bolt hold open catch prevents a 5.56 upper from seating all the way — we were tempted to break out the Dremel just to prove a point but figured the guys at POF might not appreciate us hacking up their baby for a photo op. Still, the pinholes are in the same locations, relative to both calibers.

This similarity in size carries over to the bolt carrier group. While we've all gotten used to Armalite and DPMS pattern BCGs having a fatter forward section in order to accommodate a bigger case head, the Revolution's carrier drops straight into an AR-15 upper. The bolt won't slide into an AR-15 barrel extension, giving the first clue that POF's engineers have managed to squeeze every last drop of performance out of the available real estate.

Revolution's short-stroke piston BCG next to a DI AR15's. Although they're the same external dimensions, there's a lot of work that went into opening up that bolt face. And yes, we'd rather shoot them than clean them.

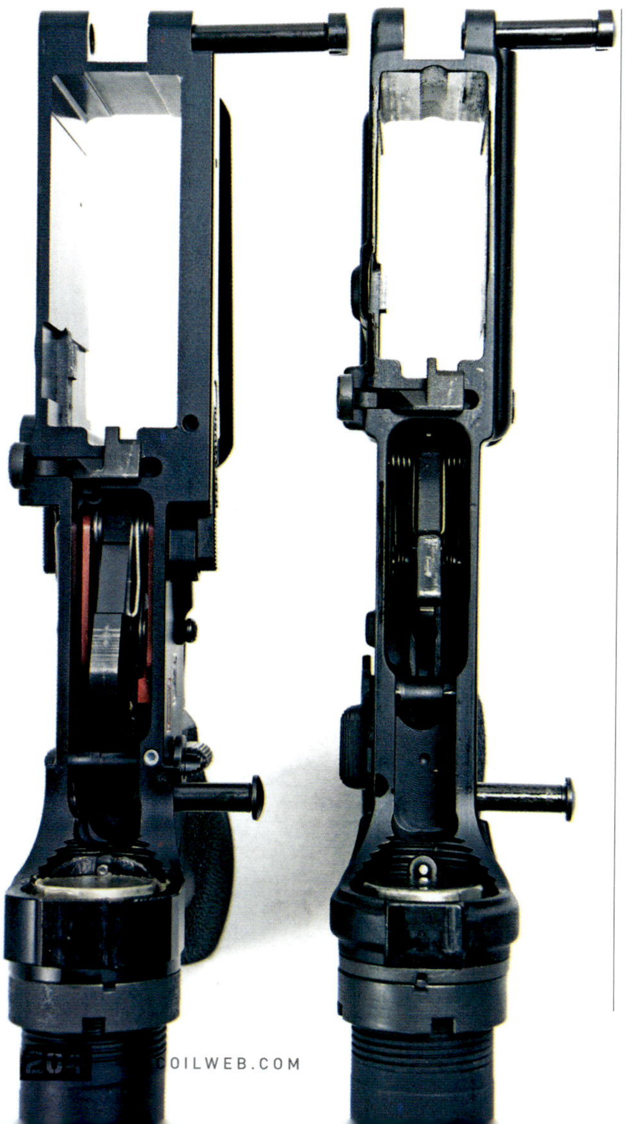

Bottom left: Viewed from above, changes to the stock AR-10 lower receiver geometry become apparent when placed next to its little brother.

These days, we're blasé that the small frame AR platform can be chambered in some pretty outrageous calibers, but there's always a limit as to how much horsepower can be wrung out of a cartridge that still fits into its magwell. There's no getting away from the facts that a 5.56 case head measures 0.378 inch in diameter, and Carpenter 158 steel has a tensile strength of 160,000 psi. Reality imposes finite limits; once you start removing material in order to fit larger shells, you eat way at both bolt lifespan and any safety margin. A x39 or Grendel case is about the widest that can safely slide into the space between the AR-15's bolt lugs, at a cost of reduced chamber pressure and life expectancy. To expect a 0.473-inch diameter .308 case running at 62,000 psi to function in Stoner's smaller design is a big ask.

SCIENCE!

To achieve the goal of fitting the larger round into a smaller space, some serious number crunching went on behind the scenes, along with the application of advances in materials science. According to our calipers, bolt lugs were made 10 thousandths of an inch wider and 5 thousandths longer. The area immediately behind the lugs grew in diameter, and a bigger extractor sneaked in, while more material was also added between the lugs. In short, every possible way to add strength back into the bolt was exploited, without increasing the overall size when mated to its barrel extension.

Finally, as stronger steel than was available in the 1950s is now a reality, a "super alloy" was selected as the billet from which the bolt would be cut. Which one? Well,

Inside the trigger guard lies the Revolution's ambi bolt hold open button. Note: QD pockets in receiver endplate.

we posed that question to POF's founder, Frank DeSomma who was understandably reticent. "Look, if someone wants to buy my gun and perform spectrographic analysis on the bolt, then I can't stop them, but I'm not going to hand over that information on a plate. I will say that we've tested it down to negative 110 degrees, and had less than 5-percent reduction in strength, which might give a clue."

Another major sticking point for the AR-10 has been the location and size of its gas port. Proof Research came out with their CAMGAS system last year, shifting the port closer to the muzzle for reduced dwell time and smoother operation, while others resort to Band-Aid solutions such as heavier buffers in order to delay unlocking through increased inertia. If the problem isn't addressed by whatever means, then premature unlocking can occur, leading to the telltale symptoms of case head swipe, difficult extraction, or, in some cases, the extractor's ripping entirely through the rim.

With an AR-15–sized BCG, the Revolution was already behind the curve with respect to the amount of inertia its reciprocating parts could apply to the unlocking conundrum, so POF took a different approach. Recognizing that adding weight to the carrier and buffer was the exact opposite tactic of that employed by three-gunners in order to reduce recoil, their engineers embraced the less-is-more philosophy, sent the carrier to fat camp, and concentrated on tuning the gas system. As a result, the Revolution is one of the smoothest-shooting .308s we've ever picked up.

Its short-stroke piston is the same as found in the rest of the company's lineup. But in this case a rifle length gas system is teamed with a 16-inch barrel, leaving a scant 2.75 inches or so between gas port and muzzle, minimizing the amount of time the gas system is pressurized and eliminating premature unlocking problems.

Further damping recoil from the 7.62 round is an effective muzzle brake, featuring three lateral ports and measuring a full inch in diameter. It's brutally efficient at redirecting gases as they emerge from the muzzle, though it increases dust signature if fired when not completely level. We tried it in brokeback prone and lost the target in the resulting haboob. Other neat touches include a fluted, medium profile barrel, encased in a 15-inch-long M-Lok handguard.

This forms part of the insurance policy the user buys in order to mitigate any risk from stuffing a .308 where a .223 should go, as there's an additional top rail machined into it which slides over the one found on the upper receiver, reinforcing this area. Its oversized, heat sink barrel nut is likewise strengthened by a solid section of handguard, giving a bit more reassurance regarding the thickness of its skinny chamber walls under the barrel extension.

Moving to the lower receiver, the user benefits from a full set of ambi controls, including bolt hold open. This is the one feature that most ARs lack, and need, as its value only becomes apparent when it's most important — i.e., that time when you're trying to get the weapon back up and running during stoppage drills. It's so much more convenient (and faster) to rack the charging handle with your support-side hand while locking the bolt to the rear without changing your master grip, something that right-handed shooters struggle with on a conventional setup. In this instance, a wide button extends into the trigger guard and is actuated by pressing upward with the trigger finger of either hand.

Filthy after its baptism of fire, it just keeps chugging along. Trijicon's 1-8 proved to be equally robust.

When it comes time to change mags, lefties gain back a little advantage. Due to its wider magwell, the Revolution's right-side magazine release button sits below the receiver's surface, and the fence surrounding it lacks the smooth radius found on the AR-15. It works fine with nekkid digits, but it's a little tight when wearing gloves and could use opening up slightly. Lighting off primers is taken care of by POF's drop-in, single-stage trigger, which is non-adjustable and breaks cleanly at 4 pounds, 8 ounces, with a short, tactile reset.

ROUNDS DOWNRANGE

As this was a .308 masquerading as an AR-15, we decided to take it up on its pretentions and drag its ass to a carbine class. Dan Brokos (featured elsewhere in this issue) invited us along to a vehicle and urban CQB course he was teaching to the fine gentlemen of the Statesville, North Carolina, police department's Emergency Response Team, so we elected to take our test gun with us. The testing protocol was simple; everything the SWAT team was doing with their patrol carbines (which ranged from privately purchased 16-inch guns to select-fire 10.5's), we'd attempt to do with a 7.62x51, shooting 175-grain M118LR. Game on.

Not wishing to be handicapped by optics, Trijicon's 1-8x28 Accupower scope was added to the mix, along with an older Leupold Deltapoint in a Burkett offset mount. Rounds were fed from either PMAGs or Lancer 20 rounders, and a SureFire X300 occupied the 12 o'clock position for the low-light portion of the class.

For five days, the class shot from every position you could think of, from, through, under and over vehicles, around cover, through glass, and from 100 yards to muzzle contact distance. We applied oil at the start of the course, then the gun wasn't touched for maintenance until it was broken down in the

PATRIOT ORDNANCE FACTORY REVOLUTION

CALIBER: 7.62 NATO
MAGAZINE CAPACITY: 20 rounds
BARREL LENGTH: 16 inches
OVERALL LENGTH: 34.75 inches
WEIGHT: 7.3 pounds
MSRP: $2,700
URL: pof-usa.com

FEATURED ACCESSORIES:
PRS Atlas Bipod BT46 - $320
Trijicon AccuPower 1-8x28 - $1,700
SureFire X300U-A - $399

PRICE AS FEATURED: $5,119

studio for this article. We ran an entire case of ammo through it, then added a few boxes of various flavors and bullet weights when we got home, just for good measure. These ranged from 150-grain factory loads to our favorite 110-grain Nosler coyote destroyers, handloaded over 48 grains of IMR 8208 XBR.

M118LR produced sub-MOA groups. We had two stoppages, both of which were failures to strip the first round from a fully loaded magazine, resulting in the case partially entering the chamber — after this, we downloaded to 19 rounds and had no other issues. The Revolution kept up with guys running 55-grain ammo and in Brokos' 10 shot Stance Check drill, allowed us to keep all rounds in the A zone at 10 yards with an elapsed time of 2.18 seconds. Figure 0.25 for reaction time, 0.3 to bring the gun up from low ready and find the sights, then we're in the realm of 0.16 splits, which for a battle rifle is pretty damn good. What's not so good is the effect on everyone around you. Behind the gun is a great place to be; either side isn't. When using the rear of a Toyota for cover, we blew out the taillights with its muzzle blast. 5.56 carbines made signature carbon prints on the hood when shooting junkyard prone. The Revolution made divots.

We'd usually prefer mounting a light to the right of the POF's four-position gas plug, but got lazy and stuck it up top in the Pic rail. It worked just fine and didn't obscure the scope.

We think there's about a ½ pound of fat still to be trimmed from the Revolution, be it from the barrel or upper receiver. At 7.3 pounds, the 16-inch version still has a little way to go before it can truly claim to be a 10 in 15 clothing, but in the meantime, it's an absolute blast to shoot. We're looking forward to picking up a variant with a 14.5-inch barrel and shorter handguard, which at a hair over 7 pounds will no doubt bring a smile to even the most jaded user. While its non-NFA, pinned-and-welded status prevents us from adding a can, it'll be fun to shoot at a three-gun match. At least, it'll be fun for us. For the rest of the squad, not so much. R

7.62 FULL METAL DEATHMATCH

Three AR-10 Carbines Go Head to Head to Head

By Iain Harrison

Photos by Kenda Lenseigne

7

Heavy

Thanks to the Salesman-in-Chief, it seems like everyone's got an AR-15. With a market awash in 5.56mm rifles, the next item many people will consider adding to their collection is its big brother, a 7.62 NATO version — because although the ammo is heavier and more expensive, by comparison it hits like the hand of an angry god. With that in mind, we assembled three of the best examples of the breed and compared them side by side to see which one might best fit your requirements.

DANIEL DEFENSE 7.62

First up is the newest of the three offerings, which may or may not be on sale by the time you read this. We got our hands on an early prototype DD7.62, so some of its features may change before its release to the public.

With that caveat in place, what's not going to change is the way the barrel mates to the upper receiver, which is both patented and unique in the AR world. Using threaded steel inserts in the upper, a full-profile flange around the barrel extension is sandwiched by the rail system and clamped by four grade-8 bolts that pass through bosses on the handguard. By doing so, its bearing surface is increased by more than 40 percent compared to a regular barrel nut, and the barrel/upper interface is massively beefed up. The weakest part of that subassembly now becomes its strongest, reducing any influence that pressure on the handguard may have on accuracy.

There's a common misconception that simply adding a free-float tube completely removes lateral force on the barrel caused by

loading the handguard. While it absolutely does reduce it by orders of magnitude in comparison to handguards that attach to the front sight tower, there's still pressure applied to the barrel nut. When our goal is repeatable accuracy out to the limit of the weapon system, every little bit counts. In the event of a barrel being shot out, or the user wishing to change calibers (and yes, we asked about a .260 Rem version and were rewarded with mumbled evasions and glances at shoes, so read into that what you will), the user simply grabs an Allen wrench and swaps the old one out. No vise block or action wrench required.

Daniel Defense has a reputation for following the AR-15 Mil-spec very closely, even when that specification has been superseded by more modern materials and processes.

Only the DD308 retains the AR-15's forward assist, but all have ambi bolt releases and safeties.

In the case of the large-frame rifle, there is no Mil-spec. Freed from the diktats of an outdated document, they came up with some very neat and original touches to improve performance. Take the bolt carrier group as an example — rather than the usual dark gray parkerizing called for by the government, they opted for a proprietary finish that offers better lubricity and corrosion resistance. The bolt now sports twin ejectors for reasons of redundancy and longevity, and has a larger extractor than other designs. Both the firing pin retaining pin and the extractor pivot pin are now captive — two fewer things to lose in the field during routine cleaning and maintenance.

Ambidextrous controls will be appreciated by lefties — mag release, safety, and bolt catch are reachable with either hand. It should be noted though that the extra right-side lever above the mag catch functions as a release only — if you want to lock the bolt back you'll still have to use the traditional method. The charging handle is decidedly non-traditional, however, as it ditches the usual spring-loaded catch in favor of a centrally located ball detent. Replaceable "wings" allow the user to choose small or large engagement surfaces, providing four configuration options.

SEEKINS PRECISION SP10

Seekins started out in the gun industry by making bottom metal and scope rings for precision rifles before branching out into ARs. It's only natural, then, for them to incorporate both aspects of the business into their latest venture, the SP10. Taking cues

Steiner's 3-15x50 in a GDI mount was used to engage all manner of targets, but we could probably have used it as a prybar or hammer as well.

from the proven iRMT upper receiver found in their small-frame guns, the SP10 features a massively reinforced barrel nut area that extends the Picatinny rail forward by four slots — as a bonus, the usual no go spot for mounting optics is eliminated. The 12 o'clock rail continues out 15 inches toward the muzzle and is triangular in profile, with a flat bottom surface to aid in shooting from packs or barricades. It takes some getting used to when used offhand — compared to the AR-15 version, this one is 3⁄16-inch wider. While that doesn't sound like much, it's enough to make us wish our digits were a bit longer.

An 18-inch stainless barrel stands out in contrast to the matte black surfaces on the rest of the rifle, complemented by a stainless ATC muzzle brake. Seekins did a good job profiling this tube, giving it a gradual taper up to the adjustable gas block and leaving plenty of material close to the chamber area. In the two years of development work that went into this rifle, it's fair to assume that time was invested in the Goldilocks gas system seen here. Evidently, a mid-length tube was found to be too short and rifle too long, but one lying between the two was just right.

Not content to leave well enough alone, the recoil system was the next target of attention. According to head honcho Glen Seekins, in order to use a regular AR-15 buffer tube, the company used high-speed video to determine the optimum velocity of the bolt carrier group, settling on a 42-coil flat wire spring, rather than the usual 37-coil round wire version. This results

In just a few short years, adoption of DPMS-pattern mags has become industry standard. Even Armalite is in on the act.

in 7.5 pounds of force applied to the closed BCG, versus the usual 3.5 pounds, making for a smoother, more balanced system. "Instead of starting out with a fast-moving BCG and then trying to slow it down at the end of the stroke, we start out slower and have a more controlled cyclic rate," Seekins explains.

Other well thought-out touches include a profiled and lightened billet receiver set, a 3.5-pound CMC single stage trigger, and dimpled takedown pins. The SP10 lower is fully ambidextrous, and the bolt hold-open and control surfaces have coarse checkering throughout.

GREY GHOST PRECISION SPECTER HEAVY

Grey Ghost Precision (GGP) was spawned from the fertile minds at Tactical Tailor, in order to expand the work of their Grey Ghost Gear brand into the rifle market. The rifles themselves are built by Mega Arms (who coincidentally are just down the road in the Great Northwet) to GGP specifications. Their home base located just outside the gates of Joint Base Lewis McChord gave ample opportunity for real world testing by guys who use rifles for a living, and feedback from troops, SOF instructors, and competitive shooters was incorporated into the final design.

Based on the Mega Maten MKM upper and lower receiver set, this is the third of our triumvirate of 10's to feature a reinforced upper/barrel interface. Apparently, people are taking notice of the weak spot in Stoner's design and taking steps to address it, and we're sure the brains behind the original would have approved. The Specter's barrel is a medium contour, 416 stainless affair with a midlength gas system, finished in black nitride inside and out. We're unabashed fans of nitride coatings in precision barrels, as both corrosion

Note flat wire recoil spring and solid stainless buffer in this Seekins SP10 lower.

DD use an enlarged flange on the barrel extension and four bolts to sandwich the barrel between upper and handguard.

Bolt features enlarged extractor and two ejectors.

and erosion resistance is improved, with no degradation of accuracy. Out front, a Gemtech Jake Brake handles recoil reduction duties and allows the shooter to attach a can via its Quickmount lugs.

An octagonal 14-inch-long, free-float tube covers the barrel and is slathered in KeyMod slots, which apart from their primary purpose do a good job of shedding weight. Like the SP10, the Grey Ghost has an uninterrupted row of Picatinny rail slots on its upper surface, while the receiver pair is tastefully sculpted and rounded, with no annoying sharp edges. The only areas we'd like to see improved are the addition of an extended charging handle to the spec, along with an improved trigger. Although the supplied unit was light years ahead of a G.I. trigger group, it paled in comparison with the Geissele and CMC bang levers on the other two guns.

ROUNDS DOWNRANGE

We outfitted the rifles with a variety of glass, ranging from a Vortex 1-6x "Jerry Miculek" scope, seen on three-gun competitors' rifles from sea to shining sea, to the exotic and rarely encountered Steiner 3-15x50 Military behemoth. Using Black Hills, Barnes, and Winchester match ammo, as well as hand loads that duplicated M80 ball, we engaged targets set at 50 to 800 yards in order to assess the three candidates' capabilities in the DMR role.

Initially, we were disappointed with the groups turned in by the prototype Daniel Defense offering, none of which were under 1.5 MOA. Since that range session, we've been informed by their director of engineering that changes to the barrel finishing process, post-hammer forging, have been made and marked improvements are evident. We've shot a bunch of DD's barrels (and currently use one on dedicated three-gun rig) and see no reason to doubt their assertion, but as always — trust, but verify.

The most consistently accurate and soft shooting of the three was the SP10, with its modified recoil system soaking up energy from the .308 rounds while it slung all bullet weights into groups that hovered around an inch. The Specter shot some phenomenal sub-MOA groups with Winchester 168-grain match ammo, but hated the usually dead-nuts-accurate Federal Gold Medal Match 175s.

So, which one would we choose? It's a tough call. Of the three, the Daniel Defense feels more lively in the hands — closer to an AR-15 than an AR-10 — but compared to the other pair, its upper and lower receivers seem less finished and

could benefit from losing an ounce or two. Lightening cuts would also spruce up its plain Jane exterior. We loved the innovative barrel nut and intelligent improvements to the bolt carrier group.

The Seekins won out in the accuracy department and had a butter-smooth recoil impulse. But while wide, flat bottoms might fit some tastes, for us they just don't cut it — we prefer something a little more rounded and petite. The rifle does have the best feature set of our lineup, however, so perhaps we can overlook its one shortcoming.

Which brings us to the Grey Ghost, which is by far the most attractive, in an inky, tubesock, post-goth kind of a way. She's no slouch in the performance department either and will probably outlast the competition due to its nitrided barrel and BCG, but could benefit from an upgrade in the trigger department.

The bottom line is that at this price point there just aren't any dogs in the group. It's like shopping for a family sedan and restricting your choices to an M5, RS7, or E63 — you really can't go wrong with any of them, all will outperform just about anything else available, but some will fit your style and needs better than others. Figure out what's most important to you, then make an educated decision based on your requirements. R

Gemtech brake directs a lot of the gas through its side ports. We wound up with a 4-inch-deep trench after this drill.

GREY GHOST PRECISION SPECTER HEAVY

CALIBER: 7.62 NATO

MAGAZINE CAPACITY: 20 rounds

BARREL LENGTH: 16 inches

OVERALL LENGTH: 35 inches (stock collapsed)

WEIGHT: 8.7 pounds

MSRP: $2,400

URL: www.greyghostprecision.com

DANIEL DEFENSE 7.62

CALIBER: 7.62 NATO

MAGAZINE CAPACITY: 20 rounds

BARREL LENGTH: 16 inches

OVERALL LENGTH: 33.5 inches (stock collapsed)

WEIGHT: 8.3 pounds

MSRP: $2,700

URL: www.danieldefense.com

SEEKINS PRECISION SP10

CALIBER: 7.62 NATO

MAGAZINE CAPACITY: 20 rounds

BARREL LENGTH: 18 inches

OVERALL LENGTH: 36 inches (stock collapsed)

WEIGHT: 9.1 pounds

MSRP: $2,489

URL: www.seekinsprecision.com

When LaRue Tactical announced that it was producing a 5.56mm AR-15–based carbine in conjunction with famed tactical trainer Chris Costa, our interest was piqued. Was this simply a clever marketing ploy to merge two respected names in order to sell more rifles, or was there something more to this collaboration? We needed to investigate further.

We spoke to Costa about his role in the rifle's creation. He pointed out that LaRue's line of OBR (Optimized Battle Rifle) carbines have been long known for their impressive accuracy and high standard of build quality. Available in 7.62 as well as 5.56 calibers, the OBR line quickly became the favorite of users of precision semiautomatic platforms. With the long-range segment covered, tactical shooters looking for the lightest build came calling. Always listening to its customers' feedback, LaRue introduced a line of ultralightweight carbines a few years ago under the PredatAR name. They are arguably the lightest full-featured complete guns of their kind produced by any manufacturer in the industry.

Shortly after Costa opened up his own training company, Costa Ludus, earlier this year, he approached LaRue about making a rifle specific to his preferences that he can run hard in his classes. We asked why he went to LaRue Tactical to collaborate on this project. Costa responded that he was impressed with LaRue's build quality and liked that the manufacturer doesn't play favorites. Every rifle that comes out of production is no different, whether it was made for him or one of his students. LaRue even had a great base gun to start with. When training is your everyday job, you learn really quickly what you like and don't like in your equipment. For Costa, it was important to keep weight to a minimum. LaRue's ultralight PredatAR weighs in at a paltry 6 pounds, 4 ounces; it seemed to fit his needs to a "T."

While Costa liked the 5.56 PredatAR's 16.1-inch lightweight barrel coupled with a 15-inch rail, he felt that the barrel from LaRue's OBR series would be a better option because of its Port Selector Technology (PST) that allows two different gas settings, for shooting either suppressed or unsuppressed, and its superior accuracy after long strings of fire. But with more accuracy comes more weight, as the OBR barrel's profile is thicker than the ultralight PredatAR's. Costa's solution was to trim down the length of the carbine to help

LARUE TACTICAL'S COSTA LUDUS PREDATOBR HYBRID

The Beard's Rifle of choice

By Recoil Staff

Photos by Henry Z. DeKuyper and Mark Fingar

Chris Costa of Costa Ludus approached LaRue Tactical about making a rifle that fit his needs as a tactical trainer. The result is the Costa Ludus PredatOBR Hybrid. Each is hand-finished in Flat Dark Earth Cerakote.

counter the additional weight of the OBR barrel. He asked LaRue to drop in the OBR barrel but shorten it to 14.5 inches with a permanently attached muzzle device for an overall length of just over 16 inches. He went on to have the already lightweight, skeletonized PredatAR handguard shortened from 15 inches to 12 inches as well. These modifications cut down some of the added weight from the heavy profile OBR barrel and ultimately made it a slightly more compact and faster-pointing gun. All of these changes produced an AR-15 weighing in at just 7 pounds, 14 ounces, an increase over the PredatAR of just over 1 pound, 10 ounces. Costa believes the supplemental weight is worth the more precise accuracy at longer range. At under eight pounds, the Costa Ludus PredatOBR Hybrid promises to be a run-and-gun carbine with top accuracy potential.

A FULL-FEATURED CARBINE

LaRue Tactical's standard PredatOBR only comes in black, with optional Flat Dark Earth (FDE) furniture. With its Costa Ludus logo engraved on the lower receiver and distinctive FDE color, the Costa Ludus Hybrid edition really stands out. Each one of the guns is individually cerakoted by hand, a time-consuming process to be sure. Cerakote is a tough-as-nails coating favored by gunsmiths for its high temperature resistance and durability. The lower receiver is machined from 7075 billet aluminum and features an extrawide, beveled magazine well for faster and more consistent magazine changes as well as an integral trigger guard. As with the OBR, the PredatOBR's barrel and barrel nut don't touch the handguard, keeping it cool and providing additional accuracy. Unlike the OBR, which features a 10 minute-of-angle (MOA) cant to its upper-receiver rail, the PredatOBR has none.

Like the standard LaRue Tactical model, the Costa Ludus PredatOBR Hybrid comes with an 1/8 twist, OBR, stainless steel barrel that boasts a life of 10,000 rounds. Costa has personally tested this, and he reports that it can be pushed beyond that limit. But we'll get to that in a bit. The carbine is also fitted with a Geissele Super Semi-Automatic (SSA) 2-stage trigger. We can't say enough about this trigger. It has a combined pull weight of 4.5 pounds, with 2.5 pounds on the first stage and 2 pounds breaking the shot on the second. The feel of the trigger is hard to beat, with a smooth-as-glass take up and a crisp break. The reset is clearly felt, and it's a breeze to get off repeated shots without jolting the gun. LaRue Tactical has also reported that buyers will have the option of choosing the Geissele Super 3 Gun (S3G) trigger instead of the SSA if they prefer.

This rifle is Costa's personal gun, set up as he runs it in his classes. His pre-production example is exactly the same as the production models, minus the Costa Ludus logo that is engraved in the lower receiver of the production version.

Like LaRue's standard PredatOBR, the Costa Ludus PredatOBR Hybrid comes with three 3-inch rail sections and two magazines (10-, 20-, or 30-round capacity depending on your location). But Costa, always wanting more, had the manufacturer load up the Costa Ludus version with even more accessories. It also comes with a choice of a permanently pinned and welded SureFire muzzle brake or flash hider. Also included were Troy Industries front-and-rear Folding Battle Sights in FDE and marked LaRue Tactical, a Magpul Compact/Type Restricted (CTR) buttstock, two Magpul PMAGs (instead of aluminum GI magazines), a Magpul MIssion ADaptable (MIAD) grip, a full set of LaRue grip adapter panels, LaRue Index Clips, and a LaRue OBR QD sling-socket adapter.

Our test model, which just happened to be Costa's personal gun, was fitted with the SureFire MB556K muzzle brake option that accepts SureFire's MINI, FA556K, or FA556-212 sounds suppressors, topped off with an Aimpoint T-1 Micro optic in LaRue's LT660 tall QD mount, as well as a SureFire M300A Mini Scout Light in Desert Sand in a LaRue LT752 Scout offset mount a SureFire SR07 remote tape switch rounds out the lighting setup. Other non-factory additions the instructor added included a Magpul Ambidextrous Sling Attachment Point (ASAP) sling plate, a Magpul Enhanced Rubber Butt-Pad, a Precision Reflex M84 Gas Buster charging handle and Knights Armament front-and-rear flip sights.

The Port Selector Technology (PST) switch located at the gas block allows for reliable suppressed or unsuppressed operation.

TIGHT TEST GROUPINGS

We took our Costa Ludus PredatOBR Hybrid out to the firing range to put it through its paces. Because the gun is designed to be a close-quarters-action carbine that can take the occasional precision shot, we evaluated it in that light. We set up a Viking Tactics-style 1-5 drill by using three International Practical Shooting Confederation (IPSC) paper targets

LaRue Tactical's 7075 billet lower receiver includes an integral trigger guard and an extrawide, beveled magazine well.

set up side by side. From a distance of 10 yards, we shot the targets from left to right, and then back from right to left during the same string of fire. Beginning from the left, the first target took one round, the middle took two rounds, and the right three rounds. Coming back from right to left, the middle target took four shots, then five rounds were deposited into the left. The completed string was 15 shots total. This drill helps test the speed and point ability of the carbine, as well as the trigger and overall balance.

The Costa Ludus repeatedly handled the drill like a champ, with each test shooter making positive hits on the 6x11-inch, center-of-mass "A" zone of the targets. The rifle was well balanced, although some of us thought it was a bit front heavy and felt that a slightly heavier buttstock would balance the gun even more. Even so, there was no oversway when pointing from target to target, just precise shots. Each rapid string of fire—up to five in a row—was made easier because of the barrel's weight, combined with the flat-shooting properties of the SureFire muzzle brake. Costa prefers using a muzzle brake to keep the rifle as flat-shooting as possible. He reasoned that if flash was a concern, he could attach a sound suppressor that would take care of the flash. The Geissele SSA trigger performed exceptionally well, and the reset was easy to recognize, precluding us from letting the trigger reset back to its first stage.

We then backed up and shot a few 100-yard groups on paper to evaluate the gun's accuracy potential. Let it be known that we do not consider ourselves to be the greatest of precision shooters. We gave it our best shot, and the groups, while competent for us,

SureFire's MINI sound suppressor is a lightweight solution and perfect fit on the permanently attached SureFire MB556K muzzle brake.

didn't feel like they showed off the rifle's true potential. Fortunately, we just happened to have Costa himself on hand to show us what the Costa Ludus could do. At 100 yards with the 14.5-inch OBR barrel, Costa was consistently shooting one-half-inch-sized, five-shot groups. He asserted that with a magnified optic, he'd been able to shoot one-quarter-inch groups before.

For ammunition, we used Federal XM193 55-grain FMJ for the entire evaluation. We suspect that match ammo would have made the group even tighter. While shooting them, Costa explained that he had performed a routine longevity test for LaRue shooting a couple OBR barrels until they were completely blown out. He reported that even after 15,800 rounds from one barrel, the carbine was still shooting a group size of 2.5 inches at 100 yards. That's pretty impressive...and not just Costa's shooting!

At an MSRP of $2,850, the Costa Ludus PredatOBR Hybrid package isn't for the feint of heart. But considering everything that's included, it's actually a pretty good value for such a high-quality, do-everything build that is packed with top-notch accessories. This gun is for those looking for a "turnkey" rifle, as Costa put it; all you need to do is add an optic of your choice. For anyone aspiring to what the real guys use, this is it. Built exactly to Costa's specifications, the first batch of these carbines are numbered, a nice touch for collectors.

So, is this collaboration between Costa and LaRue Tactical an exercise in marketing? Not after what we experienced. The Costa Ludus PredatOBR Hybrid is an exceptional setup for any serious shooter. And don't just take our word for it. If you happen to take a course with Costa (a three-day carbine course is $665), ask him to try one out. After all, he carries a spare for students to do their own tests and evaluations with. **R**

LARUE TACTICAL COSTA LUDUS PREDATOBR HYBRID

CALIBER: 5.56x45mm NATO

MAGAZINE CAPACITY: 30 rounds

BARREL LENGTH: 14.5 inches
(16.1 inches overall with pinned and welded muzzle device)

OVERALL LENGTH: 31.625 inches (stock collapsed)

WEIGHT: 8 pounds, 15 ounces

MSRP: $2,850

URL: www.laruetactical.com, www.costaludus.com

FEATURED ACCESSORIES:
Aimpoint Micro T-1 (2 MOA) Optic ($731)
• LaRue Tactical LT660 Aimpoint Micro Mount ($98)
• LaRue Tactical HandStop ($35)
• Surefire M300A Mini Scout Light ($345)
• Surefire SR07 Remote Tape Switch ($80)
• LaRue Tactical LT752 Scout Offset Mount ($98)
• Magpul ASAP Sling Plate ($30)
• Magpul Enhanced Rubber Butt-Pad ($20)
• Knights Armament Front Flip Up Sight ($190)
• Knights Armament Micro Flip Up Rear 300 Meter Sight ($125)
• Precision Reflex M84 Gas Buster Charging Handle ($90)

PRICE AS FEATURED: $5,119

COSTA LUDUS PREDATOBR

Now that we've looked at Costa's personal carbine, which is slightly modified, let's have a look at one that isn't altered, used, and abused. The production version of LaRue Tactical's Costa Ludus PredatOBR Hybrid is packed to the gills with accessories and upgrades right out of the box. Here are a few highlights.

1. The production version of the Costa Ludus PredatOBR Hybrid comes with an impressive array of accessories as well as a shipping case (not shown).

2. The Costa Ludus logo is engraved on the left side of the lower receiver with the first 100 rifles featuring individually lasered numbering.

3. The rifle comes with your choice of a permanently pinned and welded SureFire muzzle brake or flash hider.

4. Front and rear Troy Industries Folding Battle Sights in FDE are laser-etched with LaRue Tactical's logo.

5. The entire gun is color-matched in FDE and features LaRue Tactical's logo on the right side of the lower receiver. You also have a choice of two great trigger options: Geissele's two-stage Super Semi-Automatic (SSA) or its hybrid Super 3 Gun (S3G).

BE VEWY, VEWY, QUIET...

Humane. Free Range. Organic. Mil-Spec Gear. And a .308 Colt. All Good Words to Associate with Delicious Meat Products.

By Iain Harrison

Photography by Henry Z. De Kuyper

What if you also got the chance to associate words like AR-15, NVGs, and hollow point, with bacon, barbeque, and pork? Sounds like a winner, right? Texas is currently overrun with porcine invaders, and we got the chance to turn a few of them into dinner, putting together the ultimate hog-hunting kit in the process.

Over the past few years, we've had the chance to hunt hogs with everything from .22 LR to .50 BMG, and in the process have come to respect the lowly swine for its ability to survive just about anything mankind can throw at it. While you can legally shoot them with any weapon, it's morally incumbent on us to ensure that Mr. Pig goes from doing what he does best to assuming ambient temperature in as short a timeframe as possible.

Shot placement is critical. And it becomes even more critical as our caliber and muzzle energy decrease. Because we oftentimes can't ensure a perfect shot, particularly at night, it can pay dividends to step up to a bigger bullet than a 5.56mm platform can launch. While we certainly wouldn't feel under-gunned with a .223, a bigger bore allows more margin of error on the shooter's part and allows us to take shots that we might otherwise pass up. If you decide to go hog hunting with a 5.56mm AR, we highly recommend using a tough, premium bullet. The 62-grain Barnes TSX — available in the Barnes Vor-TX line or from Double Tap Ammunition — has always performed very well for us and produces outstanding results. For this trip, though, we wanted to see if the .30 caliber would give enough of an advantage to warrant the additional weight, expense, and recoil.

HOG GUNS, HOG CALIBERS

The .300 Blackout is still riding a wave of popularity, and with good reason, so its inclusion in our battery for this trip was a no-brainer. Its ballistics are often compared to the Soviet cartridge it was meant to surpass, and for years the 7.62x39mm was the only game in town if you wanted a .30-caliber AR-15. Those prepared to step up to the AR-10 platform could avail themselves of the classic 7.62mm NATO round. We were, so we did. Rather than give the airline an opportunity to lose three separate rifles on the way to and from the hunt, we built the system around Colt's clever LE901, which allows the user to attach an AR-15 upper to an AR-10-style lower. This meant we only took one serialized lower and also gave us spare uppers to share with the new friends we made in camp.

As most of the hunt was going to be under low-light/no-light conditions, the optics package was selected in order to allow us to acquire our targets under the widest range of conditions possible. We also wanted to span the gamut of weapon weights, from a lightweight 7.62x39mm build all the way to a chunky .308 Win, with a similar range of costs.

The low end of the scale was occupied by a build that started out as an AR in the Russian chambering — a Bushmasterkov, if you will. The Bushy's standard handguards were ditched in favor of a 15-inch Troy unit, which required a VLTOR low-profile gas block in order to slide over the barrel. An optional rail section was added to the business end, to which was fitted a 500-lumen Surefire X300 Ultra weapon light. In keeping with the lightweight theme, we chose a Leupold Mark AR MOD 1 scope as the day optic. In terms of clarity, this scope gives up nothing to the more expensive models in their lineup, but it does sacrifice a little adjustability and light gathering in return for lower cost and weight. In order to run this upper at night without alerting pigs until after the first shot

[Just in case you run out of ammo...]

Sixteen inches of pure nasty. The LE901-S with Bushnell 1-8.5 scope and CTC Defense laser designator.

was fired, we mounted a Crimson Trace CMR-201 IR laser under the main optic. This kept it out of the way and ensured the lens didn't become fogged with powder residue.

The mid-weight contender was our long-term test upper from Faxon Firearms, featured in the last issue of RECOIL. This time, we simply swapped out the .223 barrel and called it good. In case anyone missed Issue 6, the ARAK-21 is a long-stroke piston system with interchangeable barrels, blending some of the best features of both the AR and AK platforms. By undoing six screws, the user is able to exchange the standard .223 barrel for a .300 BLK tube, giving a decent jump up in terminal ballistics. Optics on this upper consisted of a Trijicon SRS red dot and a Laser Devices DBAL-D^2 laser and illuminator.

HEART OF THE SYSTEM: COLT'S LE901 MODULAR CARBINE

Tipping the scales at a robust 11.5 pounds, fully kitted out, was the Colt LE901. Its modular lower receiver is a pretty slick feature. By moving the typical location of the front take-down pin, Colt's engineers

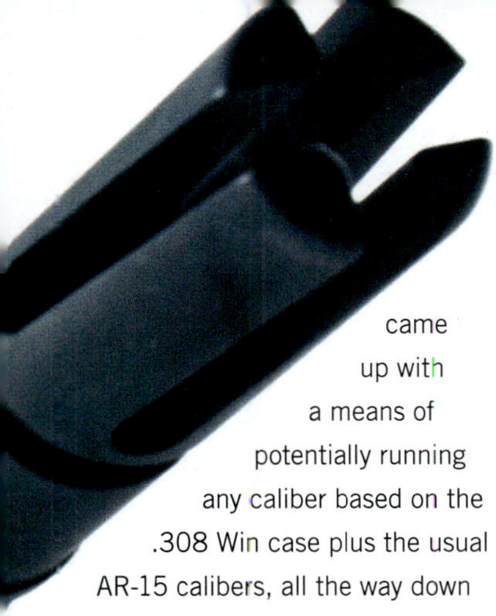

came up with a means of potentially running any caliber based on the .308 Win case plus the usual AR-15 calibers, all the way down to .22 LR, in one rifle. The key to this ability is the aluminum conversion cassette that attaches to an AR-15 upper in order to interface with the 901 lower. In our humble opinion, this is probably the most versatile AR currently available. We look forward to turning it into an SBR and making it even more adaptable.

The rest of the carbine shows evidence of the upgrades Colt has made to its lineup over the past few years. Since losing the lucrative M4 contract, the Hartford, Connecticut-based company has had to listen to the needs of its commercial customers and the result has been a more varied and we daresay better product line because of it. The 901's 16-inch, 1:12-twist barrel is MP-inspected, chrome-lined and fitted with a Vortex flash hider — a very useful addition when hunting at night using NVGs. While visiting the Colt plant last year, we saw the gun's monolithic upper receiver being hogged out of a very substantial forging by brand-new CNC machines. Start with a big chunk of aluminum, turn 75 percent of it into chips, and anodize the rest. Works for us.

Inside that monolithic upper rides what looks like a fairly typical AR-10-style bolt and carrier — until you examine it closely. The carrier's rear has a couple of raised anti-tilt pads machined into it, and next to the cam pin cutout in the upper, Colt has added a tiny steel plate. Both of these modifications are usually present in a well-sorted piston AR — any bets as to what might be introduced at SHOT Show 2014?

Round Up: 7.62x51mm, 7.62x39mm, 300 BLK — all worked well, but bigger is definitely better.

Although we didn't make use of the backup irons on this hunt, the 901 comes with a nicely machined folding front sight and a quality rear that clamps to the full length Picatinny rail. It's just as well that the rear sight leaves nothing to be desired, as a typical AR-15 back-up iron sight is at the wrong height to be of any use. Other quality touches include the quick-detach swivel pockets machined into the handguards at four locations, which match the two in the VLTOR collapsible stock.

Making best use of the cartridge's long-range potential required a versatile optic, so we fitted a Bushnell 1-8.5x24 scope in a GDI mount, which lined up perfectly with the Trijicon thermal unit up top. Down below, we used a CTC Defense MVF-600. This being an AR, we couldn't leave it in its stock form, so a few simple modifications were made to the lower. First to be installed was a JP trigger kit — with the duty hammer spring, it broke at a crisp 5 pounds and had a nice short reset. We added an Ergo grip for better hand placement and a larger trigger guard from Tactical Machining, in case the evening temperatures required the use of gloves.

AMMUNITION

With .300 BLK ammo being rarer than an honest politician, hand-loading was the best option to get the terminal performance we needed for this trip. An evening was spent cutting down and forming a bunch of .223 cases, which were then stuffed with H110 powder and topped with a Barnes 110-grain TSX bullet, developed specifically for this caliber.

In 7.62x39mm, we picked the same bullet weight, but our original plan of using the same Barnes bullet was thwarted by its length — the manufacturer deliberately made it long and skinny to aid feeding in the Blackout, which made it too long for the 7.62x39mm — so instead we selected Sierra's stubby hollow point.

Choices in .308 Win were easier to come by. Federal's MSR round with its bonded Fusion bullet is designed specifically for use in ARs, so that was a no-brainer. We also selected Hornady's 150-grain SST for its explosive expansion and huge wound channel. Although it will usually not penetrate both sides of the animal, hogs are notorious for leaving almost no blood trails, so instead of picking a bullet that might result in a pass-through, we opted for one that turns innards to jelly.

AR-15 magwell adapter.

1

2

3

1 We made a few improvements to the stock gun, adding an Ergo grip, JP Enterprises trigger, and Tactical Machining trigger guard.

2 X39 upper ready to be fitted to LE901 lower. Note: The magwell adapter pin is flush.

3 LE901 with AR-15 magwell adapter in place.

OWNING THE NIGHT

Hogs are nocturnal by nature, and as soon as bullets start flying during the daytime, they become active almost exclusively at night. In order to maximize your chances of literally bringing home the bacon, a prospective hunter should be prepared to forgo sleep and adapt to the pigs' routine. In Texas, a lot of hunting takes place around feeders, which operate on a timer to disperse corn at preset intervals and have low-intensity light pooled around them. In this case, a weapon-mounted light and a half-decent low-power scope are all that's needed to line up a shot. This being a RECOIL hunt, we decided instead to go the whole hog, so to speak, and reached out to some friends in the industry for some mil-spec night vision gear, opening up a whole 'nuther set of hunting opportunities.

For the complete rundown on image intensifiers and how to employ them, check out RECOIL Issue 4. For this trip, however, we added capability in another part of the electromagnetic spectrum — thermal. FLIR has been equipping US warfighters with the world's best thermal imaging systems for several decades, and they've developed products for land, airborne, and maritime needs. If you've ever seen footage from the sandbox of Apache helicopter pilots at work, you've seen FLIR systems in action. We had the opportunity to use two of their products, the T60 clip-on sight and the PS32 thermal camera.

If one PVS-14 is good, then two must be better. That was the thinking behind I² Technologies twin helmet mount that allows the user true binocular vision from a helmet-mounted system. Once the target has been identified, the user can either use an IR laser to paint it with an aiming mark and engage it quickly, or detach one tube from his helmet and snap it into place behind the day optic on his weapon for a more precise shot. It's a nifty idea that has been well received by certain elements of the military who wear beards. On our hunt, it also gave us the ability to share night-vision capability with a buddy, as one of us was able to run a head-mounted NVG and toss the spare to a friend to mount behind the SRS red dot.

The beauty of all this gear is that it allows us to cover more ground more effectively than might otherwise be the case. Our tactics revolved around spotting thermal signatures at distance with the handheld PS32 while rolling slowly down ranch roads in the back of a truck. Once a bunch of white dots came into the viewfinder, the two or three guys with whom I was hunting would dismount and close the distance to positively identify the targets. At about 600 meters, cows and pigs start to look a lot alike, and a full body mount of a Black Angus was not in the budget for this trip. Stalking closer in a single file, we'd move in using the

1 Running twin PVS-14s in quick-detach mounts allowed for maximum flexibility.

2 Thermal clip on scope mounts in front of the day optic, with no change of zero.

3 The I² Technologies PVS-14 has been hardened to withstand 7.62 recoil and can be mounted behind a red dot with NVG settings.

4 PS-32 standalone imager is lightweight and handy.

ENVIRONMENTAL IMPACT OF PIGS:

BIRDS AREN'T THE ONLY ONES WHO ARE ANGRY

Pigs are a non-native, invasive species that cause billions of dollars of damage every year to farmland, crops, domesticated animals, and native habitat. Originally introduced by early settlers, the problem has increased steadily since the 1880s. Currently, the U.S. population of wild swine stands at about 4 million, distributed across 40 states.

Under ideal conditions where there is plenty of food and water, such as your typical suburban development, female pigs can reproduce at 6 months of age and produce two litters of 12 piglets per year. Being very protective of their young and lacking natural predators, the piglet survival rate is dependent only on food availability and disease, barring the occasional encounter with either a bullet or the front end of a truck.

Biologists at the Texas Parks and Wildlife Department report that wild hogs can cause mass destruction when left unchecked: "Rooting and trampling activity for food can damage agricultural crops, fields, and livestock feeding and watering facilities. They also destabilize wetland areas, springs, creeks, and tanks by excessive rooting and wallowing. In addition to habitat destruction and alteration, hogs can destroy forestry plantings and damage trees. While not active predators, wild hogs may prey on fawns, young lambs, and kid goats. If the opportunity arises, they may also destroy and consume eggs of ground nesting birds, such as turkeys and quail."

Want to do something good for the environment? Go kill a pig.

Bushnell Elite Tactical scope held up well to the dusty conditions and mated with the T60 thermal sight at the correct height. Mount is a GDI P-ROM L 34mm.

wind to carry our scent away from the pigs' sensitive noses. Once we judged that they were in range, we spread out and took a knee. The sense of anticipation as the prey moves into the killing area is one of those experiences that transcends both geography and history. Mankind's earliest ancestors would have recognized the feeling, even if they would have been utterly bewildered by the equipment we carried.

In one instance, we spotted the thermal signature of a group of pigs and then made a big loop around to use the wind to our advantage, getting as close as we dared. Freezing in position as about a dozen hogs snuffled closer, we whispered our target selections and opened up, dropping one animal at only 15 yards from the muzzle. In some cases, we stayed blacked out. In others, our first couple of shots were taken using

COLT DEFENSE LE901-16S

CALIBER: 7.62 NATO
MAGAZINE CAPACITY: 20 rounds
BARREL LENGTH: 16 inches
OVERALL LENGTH: 34 inches
WEIGHT: 9.4 pounds
MSRP: $2,129
URL: www.coltsmfg.com

FEATURED ACCESSORIES:
Bushnell Elite Tactical 1-8.5x scope - $2,149
CTC Defense MVF-600 - $2,199
Ergo Grip Deluxe Grip - $41
GDI P-ROM L 34mm mount - $565
JP Enterprises Original Fire Control Kit - $120
Tactical Machining Trigger Guard - $15
Magpul Pmag - $20

PRICE AS FEATURED: $7,238

ADDITIONAL ACCESSORIES:
Faxon Firearms ARAK-21 300BLK upper - $1,200
Laser Devices DBAL-D2 - $1,845
Trijicon SRS - $1,125

PRICE AS FEATURED: $4,170

ADDITIONAL ACCESSORIES:
Bushmaster Upper - $608
Troy Alpha Rail - $209
Crimson Trace CMR 201 IR - $269
Leupold Mark Ar 1.5-4x20 Firedot - $500
Leupold Mark 2 IMS - $152

PRICE AS FEATURED: $1,738

IR lasers and thermal to target the biggest hogs, and as soon as the hammer dropped, everyone hit their weapon lights and opened up on the remainder as they charged for cover.

So which of our three set-ups proved to be the best? The 7.62x39mm and the .300 BLK both tied for effectiveness against four-footed pests. We noticed a definite improvement in terminal effectiveness over the 5.56mm at the distances encountered, and when the action got hot and heavy. In particular, the 110-grain Barnes bullet developed specifically for the Blackout was pretty much a one-shot deal. However, the hog-killing crown goes to the .308 Win, as although it was heavier, didn't swing as fast and was noticeably slower on follow-up shots, it just kills things deader from farther away. End of argument.

Our hog hunting trip was arranged via NRA Outdoors and took place on the Spike Box Ranch in Benjamin, Texas. It is a working cattle farm, and the ranch managers are only too happy to see their guests put a hurting on the pig population. While they do charge for the privilege, the payoff is good bunkhouse accommodation, great home-cooked food, and camaraderie around the campfire. Over the course of two nights, the combination of semi-auto rifles, thermal imaging, NVGs, and IR lasers accounted for 13 hogs. Just as importantly, cigars were smoked, whiskey consumed, lies told, and new friendships made. Oh, and who says you can't use an AR for hunting? R

Middle: We fitted the Faxon with a DBAL-D², which allowed both IR targeting and illumination.

RECOIL MANAGEMENT

A Former Navy SEAL Explains Why Proper Technique is More Important Than Kit When it Comes to Faster Follow-Up Shots

By Jeff Gonzales

Certain techniques look strong, but from a recoil management point of view they are — how should we put this? — less strong. Let's go back to the basics and examine the human body itself. It's critical that we utilize techniques that employ as much of the muscular chain as possible.

This is not a new topic. Folks have sought ways to improve recoil management since we all started shooting guns. The truth is that most of the time, we discover an accessory as often as we find a new technique — and they are definitely not the same. I am not a fan of accessories for a lot of reasons, but the biggest one is that they have a tendency to mask poor technique.

In my line of work, I have the pleasure of working with so many different students, each with a variety of different skills. One thing that ties them all together, aside from their thirst for knowledge, is the training scars that they bring to the class. I had one student a while back who fundamentally had good technique and was meeting all our milestones with his tricked-out rifle. For one evolution, I had him shoot my

8

Personalities

rifle for a graded drill. It was a form of stress — pulling someone out of their comfort zone. He had no doubt in his mind that the rifle was zeroed, accurate as hell and, while slightly different, still a standard AR platform. He shot the drill and did well. I asked him what he had learned through it all, and his comment was that he had to work incredibly hard on recoil management.

I don't have accessories or bling on my rifle to compensate for technique, and as a police officer who usually shouldered a duty weapon, he couldn't either. Give that some thought. You've probably heard the saying that it's not about the arrow, it's always the Indian.

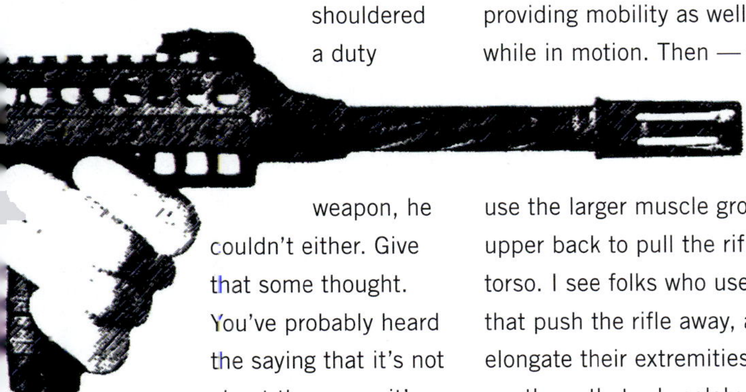

NO EXCUSES

Unfortunately, there are no shortcuts in this world. There is only hard work, determination, and inspiration. That hard work comes in the form of understanding how the body works as it relates to combat shooting. It's important to recruit as many of the large muscle groups as possible. I see many techniques that are counterintuitive to this principle. They might work for the short duration of stepping up to the line to perform a drill, but will they work when you have been on patrol all night, or had to hump all day to the target, or are just getting over an illness and not at 100 percent?

I like looking to the world of heavyweight lifting, particularly overhead lifts, as a good reminder of how important structural support — specifically skeletal support — is. If the skeleton is not properly aligned, you are simply not going to be able to support heavy loads. If you drop down in weight, you get away with plenty of poor lifting techniques. But the moment you start to get some serious weight on the bar, it's a whole 'nuther ball game. You either employ solid form and perform well, or use poor form and get hurt.

STRUCTURAL SUPPORT

Your technique should take advantage of a stable, athletic base. It should be athletic in the sense of providing mobility as well as stability while in motion. Then — and this is the key — you have to use the larger muscle groups in the upper back to pull the rifle into your torso. I see folks who use techniques that push the rifle away, as they elongate their extremities. Then there are those that rely solely on their arms to pull the rifle. The shoulder is an amazing ball joint; it can roll forward, rearward, up, and down. You are much better with a rearward and down posture as opposed to a forward and up posture. We see so many folks rounding their shoulders forward. I grew up in Texas with a mother who was constantly telling me to sit up straight. That meant pull your shoulders back and stick your chest out — good posture. (I cannot believe I just quoted something my mom said as it relates to shooting. Who knew that'd ever happen?!)

In the end, relying on accessories over technique for recoil management will not create the best shooter. While it may seem viable at the time, an accessory is not a replacement for technique. Instead, get better at employing your God-given talents — use your strength and bring to bear your entire muscular chain. You are only as good as your weakest link. R

Regardless of ready position, the shooter should employ an athletic stance, balancing both mobility and stability.

Notice the shooter indexes his elbows toward the feet. This allows for maximum muscle recruitment and power.

As the shooter mounts the rifle, the lower body should be stabilized. The stance is completed by pulling the rifle into the shoulder pocket with arms, back, and core.

ABOUT THE AUTHOR

Jeff Gonzales is a former U.S. Navy SEAL and president of Trident Concepts, LLC, which provides clients with the full spectrum of weapons and tactics training necessary to address the challenging needs of modern warfare and personal defense. Trident Concepts' roots come from Naval Special Warfare and bring the same high-intensity mindset, operational success, and lessons learned to its training programs.

› www.tridentconcepts.com

RO

RIFLES FOR CKSTARS

556 Tactical's Deathpunch AR-15

By Aaron Cowan
Photography by Straight 8

Guns and rock go together like peanut butter and jelly. But guns and metal — they go together like steel and brass. Whether a Hollywood blockbuster or videos on YouTube, if there's film of weapons at work, it's usually set to massive sound — electric guitars, emphatic beats, and driven lyrics that reach right into your chest and motivate your blood pressure. It's a rush in the form of sound, paired with the feeling we all get when turning money into smoke one skilled trigger press at a time … or a "quick, no one's looking" bump fire (yes, we've all done that). Not all bands get involved in the gun industry, and fewer still seem to have the same passion about gun rights as Five Finger Death Punch. I've been a fan of the boys from the West Coast since I first heard them in 2005, but when FFDP guitarist Zoltan Bathory took to Facebook in 2013 to speak out on gun rights, his comments hit me right in the Second Amendment feels.

Zoltan wrote (in part), "Let's talk about 'assault' rifles for a second. Anti-gun advocates in the media placing all the emphasis on the word 'ASSAULT' for a reason. It sounds menacing to those who don't know much about weapons, and this particular word also embodies the concepts of 'first strike' and 'aggression'… and the media just loooves that. The fact IS, however — a rifle is an inanimate object with no free will or intent on its own. It doesn't plan to assault anyone … So really, this is all just playing with words … If I called my rifle a 'defensive rifle' — it wouldn't throw up the red flag — and in reality THAT'S WHAT IT IS. I have it for defensive measures so that's what it should be called. A 'defensive rifle.'"

There has been a fair share of celebrities from film and music professing to be pro-gun and "pro-gun." But they often ride the sporting purpose bandwagon and avoid getting into the more politically charged issue of firearms, misnamed as assault rifles, as tools for self-defense. Zoltan doesn't appear to be concerned with toeing some political line, and neither does any other member of FFDP. They like guns for the same reasons most of us do: self-defense, practice, sport, challenge, and the passion of the pro-gun community. The band has also made sure to take their application of all guns sleek and sexy to their videos, most notably with the video for *The House of the Rising Sun*, an excellent cover of the classic folk song. The video features guest

appearances by the Kriss Vector, Midwest Industries, Magpul, Glock, Tactical Tailor, Grey Ghost Gear, and enough Crye Precision gear to make a Zoltan action figure a viable option. All things considered, it's hard to doubt that FFDP is pro-gun and supportive of the industry as a whole, and it's this attitude that attracted the attention of Louisiana-based 556 Tactical.

Most unique rifles begin with a good story, and the rifles that would become known as the DEATHPUNCH rifles are no different. Colin Evans of 556 Tactical read a post on social media about FFDP passing through Louisiana. He contacted the band and invited them out for some range time. Because there is nothing more metal and American than range time, they accepted. It wasn't long before the 556 Tactical crew and the band became regular friends, and the idea for a band rifle was born. The guys in the head shed at the ATF allow for 11 character serial numbers on firearms, which all but begged for a unique band signature to grace each of the nine rifles to be built.

556 TACTICAL

556 Tactical started out like so many rifle companies, whereby a passion for shooting and skill at custom builds evolved into a business. Word of mouth about the quality of the 556 Tactical rifles spread, and the company has continued to grow ever since. They build AR-15 and AR-10 rifles to order, and no reasonable customization is out of the question. In speaking with Colin, it didn't take long to get the vibe for the company — a group of shooters who are serious about what they build and the customer spending their hard-earned money. But they're not really interested in taking themselves too seriously in an industry that has more "tactical" and "operator" than perhaps it really needs. They make it pretty evident that you can build custom, hard-use rifles without the usual dramatic push to have their rifles seen as the last, best hope for mankind or the first choice of elite door kickers everywhere. Not that their rifles don't meet that standard, but in a market of MultiCam-and-magazine-dump visuals, 556 Tactical builds rifles and lets them speak for themselves.

If you're going to make custom rifles, you have to live by the word "custom." If the customer wants it, they get it, and when your appeal stretches from recreational shooters to the aforementioned door-kicking community, you can't be too rigid in how an idea takes shape. So when ideas for the DEATHPUNCH rifles were being discussed, Zoltan wanted to put his personality into his rifle with a design that went beyond basic customization of the proposed limited run. Now if you don't know Zoltan, he's perhaps one of the most recognizable members of the entertainment industry. He drives race cars and runs a gun like a professional. So when you combine those things and try to put them on paper, you might get something that looks like, well, a rifle themed on an exotic supercar.

PUNCH IT

Few things are as visually iconic as the Ferrari: sleek lines, aerodynamic shapes that border on science fiction, carbon fiber that is as much functional as it is sexy … and the racing red. The Rosso Corsa red, a color that identifies Italian race cars, has been part of the Ferrari identity since the company began producing dream-car poster subjects — and prop furniture for women in bikinis since 1929. Paired with the bright yellow brake calipers found peeking out of the rims of almost all Ferrari Red supercars, you have an instantly recognizable color combination that invokes a feeling of horsepower, speed, and for most meat-eating men, *want*. You'd be hard pressed to find a Ferrari not on a short list of dream cars tucked away in all little boy's heads, and that desire sticks with most of us into adulthood. Zoltan is certainly no different. When the hard work the band put into Five Finger Death Punch gained them recognition and then fame, that Ferrari was finally within reach. When it came time to develop a rifle scheme with 556 Tactical, a Ferrari-themed rifle only made sense.

Glaring red, bright yellow, carbon fiber. It isn't a look for everyone, but then neither is rock music. Combine a metal guitarist's love for Ferrari imagery and guns, and you get the DEATHPUNCH1 — a custom rifle from stock to barrel. Built on the 556 Tactical ECR billet upper and lower, it's powered by a CMC 3.5 pound curved profile trigger, with a 14.5-inch fluted stainless barrel chambered in .223 Wylde mated with a NiB-X bolt carrier group. The barrel is encased in a 556 Tactical Evo rail that has been hydro dipped in a carbon-fiber finish, as are the Magpul grip and B5 Systems SOPMOD stock. The whole package is finished off with a 556 Tactical StrykeHyde flash hider and Samson Lo Pro sights.

Aesthetically, the carbine captures exactly the original design intent — a rifle that embodies the classic look of Enzo Ferrari's creation. With its instantly recognizable color combination, it's impossible not to look at this rifle and think of V12s and every single supercar you've ever seen in person or on TV. Do Ferraris come in other colors? Who cares? The racing red is all that matters.

If there is one thing we can agree on in the gun community, it's that we can't agree on everything. We don't have to. Be it black, OD green, Kryptek, or Tiger Stripe, a color scheme is a personal choice — it's all about personality and taste for range guns, or mission and terrain for working rifles. The DEATHPUNCH carbines show us the less-serious side of the industry. We have a band that worked hard to get where they are, and they are celebrating that by teaming up with a company that worked just as hard to get where they are. And they're bringing a few more AR-15s to life with their own unique personalities. In a market jam-packed full of quality rifles, buyers have more options than ever when looking for a home defense rifle, a collector's item, or a pure range toy. No matter the category, personalization is always going to be in demand, and 556 Tactical meets that demand well.

Only nine DEATHPUNCH rifles will be produced — number 1 is Zoltan's, and FFDP vocalist Ivan Moody laid claim to number 7 (its flat black and chrome theme is very impressive in its own right). The remaining DEATHPUNCH rifles, all autographed by Zoltan, will be sold through 556 Tactical.

Do you need one? Well, sometimes want should dictate need. Even if you aren't fortunate enough to get your hands on a DEATHPUNCH rifle, there's nothing keeping you from putting your hands on one of 556 Tactical's Stryker carbines — they get five thumbs up.

Q&A WITH ZOLTAN BATHORY OF FIVE FINGER DEATH PUNCH

RECOIL: You grew up in Hungary. How did you first get into firearms?

Zoltan Bathory: Believe it or not, I had a bright future in long-range assassinations ... but I threw it all away ... [laughs]! In communist countries, it was fairly common for the Army to come to the schools and recruit kids for future careers. I was around 12 years old when I finished in first place at their airsoft competitions, so they invited me to come and train with the snipers. I was shooting bolt-actions with them for long months, but when I got bored I'd pick off some birds that decided to fly over the range.

The captain in charge nearly lost his mind. There was a town behind the sand hill barrier, and if I'd had missed the bird, the projectile would have landed somewhere in town ... As a 12-year-old cocky kid, I didn't even consider the possibility that I could miss. And that pissed off the captain even more. Rightfully so; he fired me on the spot.

It was the ultimate lesson about discipline. I was a much better shooter than the other kids there, but that didn't matter. Anyone in their right mind prefers to train an OK shooter with good discipline than a prodigy with a bad one. I learned my lesson, but I never went back as my attention turned toward playing guitar instead. Technically I still became a hit man, just writing hit songs.

You can have pretty much any gun you want and yet chose the AR-15 as your personal canvas for customization. Why?

ZB: I like the AR-15. It can do a little bit of everything, and that versatility is what makes it the perfect all-purpose rifle. It's lightweight, it's easy to get ammo, it's easy to accessorize ... you can easily attach all sorts of stuff ... as you see I even attached a Ferrari to it.

Now why red? [smiling] Well, we customized it for safety. You see, I saw on TV when Piers Morgan and a few other "rocket scientists" — simply not knowing a thing about guns — inadvertently suggested that between two rifles of the same caliber the black ones are far more dangerous. Whoa! So that's why the evil-looking black ones are called "assault rifles." Who knew, right? Once I heard that from the experts such as Piers, I quickly locked all my black ARs in a safe because I was afraid of their bad attitude. And you know these inanimate objects may even assault somebody with their "blazing semi-auto rapid fire," which I didn't even know these guns were apparently capable of.

So I figured, if this new one is painted Ferrari Red with a lot of carbon fiber and a yellow trigger, it will behave like a polite sports rifle and it won't go and do any of that assaulting stuff! So that's how we ended up building the red rifle. Look at it, I mean how could they ban something so harmless? And don't forget, there is a tactical advantage here too: I can throw a gun rack in the Ferrari, right in the back window ... and you won't even notice the gun, 'cause technically it will be camouflage in this color combo.

You're an accomplished martial artist as well as a firearm enthusiast. Do you see the two as separate disciplines or part of a continuum?

ZB: A continuum, absolutely. With the popularity of mixed martial arts, even highly accomplished martial artists loyal to one certain discipline began to accept the fact that no single style can dominate all the others. To be complete martial artists we have to expand our horizons and incorporate every move that can answer a possible situation. But why stop there? Real life has no referees. Edged weapons and firearms are part of our reality, so it would be crazy not to train judo, jiu-jitsu, kung-fu, while considering the involvement of weapons. So I've been training in military combatives, close-quarter combat, and combat driving, because it is as real as it gets. And it's a hell of a fun thing to do. I want to go all the way to level 4 CQB instructor, and down the line I would like to build a training facility for special operators.

Following a January interview, you caught flak from both ends of the political spectrum after you qualified your support of the Second Amendment with the idea that gun registration was a good thing. For the record, we think you're wrong, but how do you think this would improve things?

ZB: Ha ha, yes, some think I'm a gun nut, some think I should take a harder line. I just can't win. But to fully understand what I actually said, you would have to put it all back into context.

I was debating with a European journalist, defending the Second Amendment, giving him multiple scenarios why it is reasonable for me to have a 30-round magazine, or how gun laws backfired in Chicago and will backfire everywhere else. It was a two-hour, ham-fisted debate that eventually got truncated to fit into an interview format. Inevitably, some things got lost in translation.

This is a very dynamic ever-evolving issue where our opinions could be situational and could be changed by a single event. For example, around the year 2000 the NRA actually supported the idea of mandatory background checks, and at the time they even had ads in the papers saying it was the reasonable way to go. Of course, certain moves on the political chessboards caused the NRA to react and change their stance. But the background check issue still remains a head scratcher.

Cold hard logic dictates that felons of violent crimes or mentally ill people shouldn't be able to obtain firearms. If you stay with that logic, however, then for this background check to be effective it would have to become universal and mandatory ... and there lies the dilemma. The background check generates a list. If you were checked out that means now "they" know you bought it, and "they" know exactly what it is you bought. So I was reasoning that the registration argument is moot, because the background check technically registered your weapon anyway, so then why don't we use it as a bargaining chip?

Politics are a give and take, and a keen understanding of which battle to pick. So I made a hypothesis, sort of thinking out loud, saying that my fellow NRA members won't like this, but I would exchange the registration for the re-enforcement of the Second Amendment — if and only IF the Second Amendment is "unf*ckwithable," irrevocable, and etched in gunmetal, then I would register my guns. It would be a win in the states that already violate the Second Amendment because they would have to lift their unconstitutional bans for this deal to happen. And we are giving up info they essentially obtain anyway through the background checks.

Now, if we don't want background checks out of our gun grab paranoia, then that's a whole new can of worms ... and also a moral question ... Are we OK with Billy Badass and Short Bus Bobby walking in the gun store with a list of people who wronged them ... and as long as they don't foam at the mouth they could buy anything they want?

You use a lot of military imagery in your videos. Is this a reflection of your fan base or do you have further aspirations beyond music and onto the big screen?

ZB: It's both. We are probably the most popular band amongst members of both military and law enforcement, because our songs often talk about values those guys understand and associate with. And we walked the walk, did a few tours [gigging] in the sandbox, and played countless "secret" shows on military bases, so we are really tight with our fans in uniform. As for the big screen, yes of course it is another medium I can use to express some ideas or fight for some causes. Right now I'm writing the treatment for our next video, which we will dedicate to PTSD and homeless vets. I want this video to hit America right in the chest. The people who love this country the most are the ones who put on the uniform and risk their lives for it, and America keeps failing these guys. The VA doesn't have the budget to take care of them — any given day, there are 300,000 homeless vets, 460,000 vets living with PTSD. And 5,000 vets commit suicide every year ... that's a veteran every two hours. That's crazy! The general public needs to know this.

You've been vocal and active supporters of serving military personnel. How and why?

ZB: It takes a special kind of person to do what they do and they have my utmost respect. We all have various jobs, but none of us have to put our lives on the line like cops, firefighters, and the military do. The general public takes them for granted and that's a shame. In fact, these guys should be the ones with big salaries since they are the ones taking on the ultimate risk. The rest of us just sleep, work, play, eat, and live not thinking this could be our last day alive. We are all domesticated and soft. But those guys, when you need, them they can't and won't run away — they will risk their lives for you. They are the last remaining lions roaming amongst us... R

BUTTSTOCK: B5 SOPMOD

BCG: NiBx

UPPER/LOWER: 556 Tactical ECR (billet matched sets)

BARRELS: 14.5-inch stainless/fluted (556 Tactical)

BARREL: 1:7/223 Wylde

MUZZLE: 556 Tactical StrykeHyde Flash hider

GAS BLOCK: Samson Lo Pro

RAILS: 12.37-inch 556 Tactical Evo

GRIP: Magpul MOE

SIGHTS: Samson Quick Flip Folding sights

TRIGGER: CMC 3.5-pound Flat

9

Cool

MAXIM DEFENSE PDX

Finally! a Rifle the T-Rex Can Actually Shoulder

By Dave Merrill

We travel nearly an hour across dry, dusty back roads in rural Idaho. Eventually we pull to a stop outside of a regular-looking steel building on a small family farm. There's nothing exceptional going on — at least not until you step inside.

This is the Maxim Defense MukWerx facility, and its remoteness ensures no one will inadvertently come across what they've been working up. Inside, more than a dozen 3D printers are churning and buzzing away on new parts and prototypes.

The old mixes with the new; a weathered manual mill up front, with a CNC machine in the next room. A box of potatoes and another of onions in front of a large rapid prototyping machine belies the rural origins.

This is where Maxim Defense made their transition from a firearms parts company into a full-scale firearm manufacturer. It shouldn't be a huge shock that Maxim Defense made this move. Remember, Magpul started by making rubber pull straps for magazines, SureFire started with bulky lasers, and Daniel Defense used to only make handguards. Insofar as the PDW-stock scene is concerned, what was once a crowded and competitive market has been largely dominated by Maxim Defense. Second place is now a long way away.

3D printers churning away. Iain may or may not be drinking beer. (He is.)

Potatoes, onions, and rapid prototyping machines.

THE SHIFT

Of course, their first rifle falls directly inline with what we know about Maxim Defense: The PDX PDW; one short enough even the classically stubby armed T-Rex can finally use. Putting together a functional production PDW requires more than just slapjacking parts together from Brownells — at least if you want to make your mark. You need the right team, and someone else inadvertently gave Michael Windfeldt of Maxim the right ones.

Timing is everything, and in this case Big Blue — Smith & Wesson gave Maxim the opportunity. As S&W moved all major operations of Gemtech Suppressors to Massachusetts, employees that didn't want to relocate across the country would be let go. And Maxim pickec them up. One of the most memorable and remarkable names included among this new staff is Dr. Phil Dater. Tinkering with, and then full-on manufacturing silencers for more than five decades, you can find Dater's DNA on many modern designs, and his fingerprints are all over the PDX.

THREE AIN'T A CROWD

No subcaliber carbine here, the Maxim PDX will be available in three calibers: 5.56, 300BLK, and somewhat of a surprise, 7.62x39. Making an AR *that actually runs* in 7.62x39 is a challenging endeavor; we've seen far more failures than successes. Triple-so when it's in short-barreled format. And with a 5.5-inch barrel? Fuggedaboutit. It may sound funny, but this is absolutely a huge accomplishment for a mass-produced rifle.

In fact, it's for this reason that we decided on this steel-case eating commie cartridge stubby carbine to base our review on instead of more traditional calibers.

FEATURES & DETAILS

Firstly, bear in mind that we're using some preproduction guns, so there'll be some relatively minor changes between what graces these pages and what you buy in the store; we'll point these out as we go along. Ours being preproduction or not, by the time you're reading this you'll be able to order a PDX from a distributor.

The stock of the PDX appears very similar to Maxim's traditional PDW stock options, but there are some significant deviations. Firstly, the upper receiver of the PDW is contoured to allow for a reduced width between the stock rods, and importantly, it brings the rods slightly higher, relative to the receiver itself. This allows for easier access to the safety selector while the stock is fully collapsed, one of our minor bitches about the legacy

Maxim Defense PDW stock.

Additionally, instead of having to use a carbine tool or some sort of thread locker to secure the short receiver extension, a simple ⅜-inch ratchet is used to torque everything down via a convenient socket hole on the butt.

Along the same lines, the Hate Brake, a gas booster and linear compensator/flash hider can also be taken down and assembled with the same ⅜-inch ratchet. After watching very high-speed video and shooting the PDX for ourselves, even with particularly flashy ammunition, we can assure you that the Hate Brake does indeed work both at hiding flash and as a compensator. Admittedly the Hate Brake was far better in the 5.56 version of the PDX than the 7.62x39. Remember, at least half of flash produced has a direct correlation with ammunition selection — and of course, the only thing consistent about Russian ammunition is inconsistency.

The Hate Brake also pumps some gas pressure back into the system, something that's important for functionality when we're talking about such short weapons. The amount of gas flow back through the barrel is about equal to the average silencer. Pop off the Hate Brake, thread on a can, and you'll more than likely

Unlike the lies you tell to potential partners, this one is indeed smaller than it looks.

be totally good to go. Also similar to some suppressors, you'll experience some amount of free bore boost. We were surprised when we were told that 7.62x39 metered at 1,900 fps; this is roughly equal to the velocity we get from our 7.75-inch Mini Draco SBR.

If further tweaking is needed, there are still some options. Sure, you can play mucky muck with different buffer weights, but you don't have to. Each PDX gas block has a threaded jet port inside it which allows one to custom tune their gas flow — it's a lot like the gas plug used in SCARs, though it's inside the gas block and not the barrel itself.

Short, suppressed guns are far more prone to gas port erosion than their full-size brethren, and the jet porting means you'll get far more life out of a barrel before it starts overgassing like crazy. The jet ports can't be swapped in five minutes in the field, but it's specifically designed that way. If you can pull a gas block and rotate a wrench, you can swap out gas jet yourself. If you can't, leave the damn thing alone.

While the preproduction models have receivers unmarked and unmarred by caliber markings, production guns will have the caliber deeply etched right on top of the forward assist.

BACKWARD COMPATIBLE

Given the unique nature of this stubby shooter, it should be no surprise that nearly all of the PDX consists of proprietary parts — from the gas system length to the bolt carrier group to the buffer system. However, if you have an M16 lower or similar, any PDX upper will drop right on once you move the buttstock assembly over to your new host. Your favorite AR trigger will also work, though since the PDX ships with a Geissele SSA-E as standard, we doubt you'll want to be playing with that anytime soon.

Something to keep in mind is that you can't shotgun the Maxim Defense PDX open like a normal carbine due to the particular buffer/BCG combination. It's not terribly hard though; just pop both takedown pins and pull the upper off from the front. Reassembly is exactly the same, but in reverse.

While the handguard of the PDX has integral sling quick-disconnect points, it's also compatible with Maxim M-Rax accessories as well as M-Lok so you aren't stuck to one location. Speaking of the handguards, you can use HK 416 compatible ones and vice-versa.

AT THE RANGE

With the very first pull of the trigger, Iain belted out a hearty laugh and was wearing a big smile. My first reaction was a mix of goose bumps and exclaiming, "Holy sh*t!"

It's not everyday shooting a new firearm can illicit that reaction from us. There's some kind of black magic voodoo going on here, of that we're sure. The felt recoil of these tiny barrels was less than that of our other, much longer SBRs. Popping plates at various ranges was effortless. Formal accuracy testing will take place once we get production models in-hand, but we're pretty sure they'll be more than simply functional.

Plugging data into our ballistic calculators, using plain-old Wolf 124-grain FMJ zeroed at 100 yards the drop at 200 yards was just around 10.5 inches, but it definitely drops precipitously from there. A precision rifle this ain't, but you can reach out further if you know your holds. So far as what a PDW should be capable of — this one is no slouch. If you want to get more into the nitty gritty of the ideal

A 3/8-inch ratchet is all you need to crank this down.

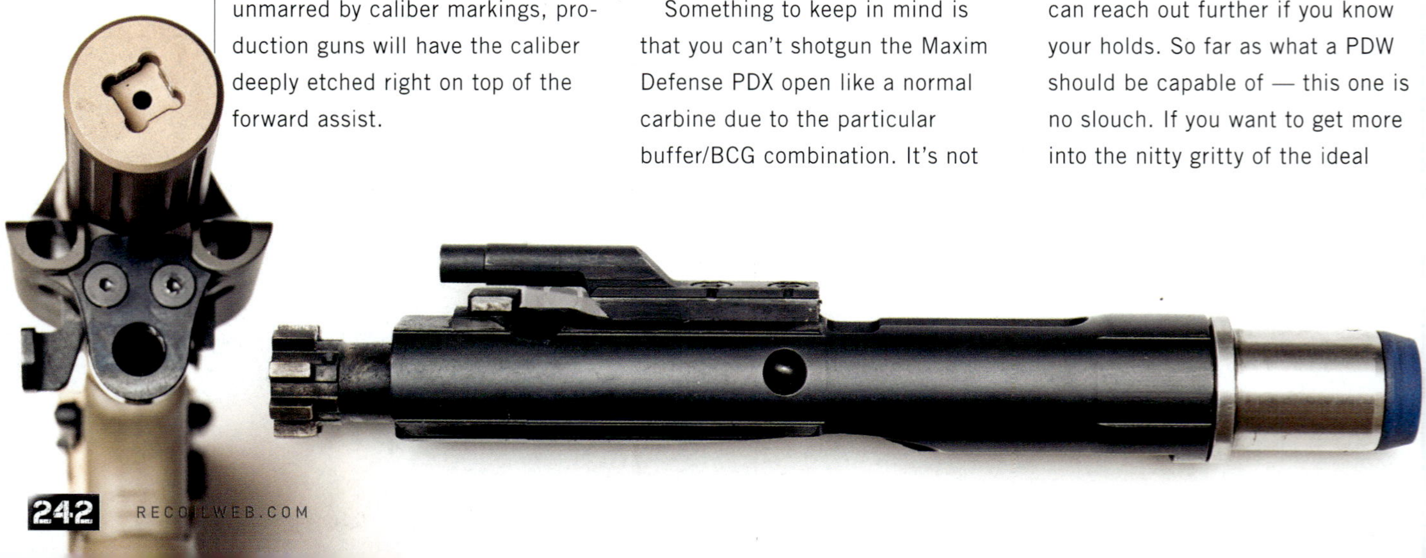

zero, see our article in Issue 35, "No True Zero."

THE FUTURE

This just the first taste of what's to come from Maxim Defense. While the 5.5-inch PDX is the premiere rifle, by the time this article hits print these will be available in both 8.5-inch and 10.3-inch configurations. This is merely the start of the Maxim Defense MDX Battle Rifle System.

But it's not just about barrel length. If you don't want a Hate Brake you can roll with something more integral: the Maxim Defense Heartbreaker silencer.

All in all, if you're looking at multiple calibers and configurations from Maxim Defense MDX Battle Rifle System, it probably exists. Let's break them all down here (and there's more than you thought)!

- Pick your caliber (7.62x29, 5.56 NATO, 300BLK)
- Barrel length (5.5, 8.5, 10.3 inches)
- Handguard
- Silencer or Brake (Hate Brake or HeartBreaker Silencer)

This all adds up to more options than we're willing to kick out in this article — and more options is more better for consumers. **R**

MAXIM DEFENSE PDX

CALIBER: 7.62x39

MAGAZINE CAPACITY: 20-30

BARREL LENGTH: 5.5 inches

OVERALL LENGTH: 19 inches

WEIGHT: 5 pounds, 11 ounces

MSRP: $2,299

URL: www.maximdefense.com

FEATURED ACCESSORIES:
Trijicon MRO - $579
Geissele Super Precision MRO Mount - $95

PRICE AS FEATURED: $2,973

ACTUAL INTEGRAL

We Dig Into the Suppressed 5.56 Integra from Gemtech

By Dave Merrill

In mid-April 2016, USSOCOM officially started solicitation for a Suppressed Upper Receiver Group (SURG). The requirements for the contract were extremely lofty and eventually the program was scuttled. Then reborn. Then got mucky again. Thus is the nature of government procurement. Several manufacturers submitted samples, but at the time of this writing, Gemtech is the only one who brought their offering to the commercial market — almost.

There are a couple of differences between the SURG upper and the Gemtech Integra, but functionally it boils down to this: The Gemtech SURG upper has just over an inch more rifled barrel, and the overall length of the upper was reduced by 1.7 inches due to government requirements.

To be sure, an integral silencer isn't a normal first NFA purchase. When starting out, buyers typically want something with more perceived utility, such as one of the myriad of multi-caliber silencers. Later, they branch out into specific calibers and guns. An integral is likely at least four or five slots down the list, if one is ever purchased at all.

SINGLE-STAMPING ACROSS STATE LINES

As an individual, before crossing state lines with most types of legally owned federally regulated firearms like machineguns and short-barreled rifles (SBR), you need an approved BATFE Form 5320.20. Though it's very dependent on your examiner, it's not unheard of for an approved form to take four to six weeks to arrive. We know numerous people who were unable to bring their goodies with them to a class or other event for lack of approval. For reasons that we don't understand and don't want to ask about, this annoying law about interstate travel applies to all federally regulated firearms, sans two categories: Any Other Weapon (AOW) and silencers.

Provided that silencers are legal in your destination state, you can cross state lines with no notice and no permission. Hence, the advantage of the one-stamp silenced rifle.

While a one-stamp rifle has certain legal benefits in the United States, there are some additional perks too. Namely, a manufacturer can build it from the ground up with complete integration in mind. The Gemtech Integra certainly isn't the

The titanium monocore is easily removed with a 3/8- inch socket wrench.

The four angled ports in the end of the monocore make the Integra self-tightening.

first gun with a built-in suppressor. In recent years we've seen several new models come to the market, such as the SilencerCo Maxim-9.

Many integral rifles have a silencer monocore machined into or permanently attached to the barrel. Some psuedo-integrals simply have a silencer welded to a shorter barrel. Both of these methods are easy ways to reach that 16-inch minimum of a non-federally regulated rifle. Gemtech took a different approach with the Integra, opting to weld a sleeve to the barrel that encompasses a removable titanium monocore. With the twist and turn of a ⅜-inch socket wrench, the core comes out for cleaning and servicing. The core itself features a sharp edge that cuts any accumulated carbon during removal. There are four angled ports machined into the end of the monocore, making it self-tightening.

Get an endcap strike with a permanently attached core and you're SOL, whereas the core on the Integra can be readily replaced. Additionally, have you ever try to run military bore patches through a silencer? We don't recommend it.

The ability to remove the entire monocore assembly leaves room for upgrades by Gemtech into the foreseeable future. Perhaps down the line, we'll see silencer upgrades for this rifle.

We admit, seeing that this was a pistol-length gas system with a monocore gave us some pause. But we withheld judgment until we got it to the range.

5.56 silencers are loud. The round's high pressure and velocity just aren't conducive to that movie-quiet experience that one can attain with subsonic .300blk or .22LR. Though in theory subsonic 5.56 could be used to further muffle the report, its terminal ballistics leaves a lot to be desired. If you ever want to see a face worthy of being featured in a sad Sarah McLachlan ASPCA commercial, we suggest observing someone who just shot their brand-new 5.56 silencer for the first time — and who has never experienced one firsthand. You can practically hear "Angel" playing in the background as the tears of sadness and regret roll down their faces.

Aside from the caliber being loud in general, the suppressed AR-15 can be unusually troublesome at times. Over the last six decades of development, we've seen a series of engineering challenges updating and upgrading the AR-15 to fit current mission requirements. These unplanned impositions ultimately made the AR the modular monster that we currently know and love. Dealing with silencers was but one of those ongoing obstacles.

By containing the expansion of hot gases at the muzzle in order to bring down the sound, a silencer

can significantly increase backpressure into the system. This can mean malfunctions from excessive bolt speed, bolts attempting to unlock far too early, more pronounced expulsion of gas out of the ejection port, and higher rates of fouling, not to mention a face full of burnt powder. After these last several paragraphs, one may get the impression that we hate suppressed ARs, but nothing could be further from the truth.

These problems all have solutions in the form of all manner of specialized adjustable parts, weights, and differing silencer designs. Modular monster indeed. The Integra does something different enough that they patented it, and now that said patent has gone through we can share it with you. It's gone largely unnoticed, mostly because it's welded to the barrel and hidden by the monocore sleeve. The Integra has a bore evacuator. Like the main gun on a tank. Here's how it works:

› The projectile first passes the primary gas port, and some of the gases travel down the gas tube to begin the cycling of the rifle as usual.

› Shortly after this, the bullet passes two additional gas ports. These ports lead into a chamber that surrounds the barrel expressly for this purpose, and a fixed amount of high-pressure gas is held.

› Once the projectile leaves the barrel, the resulting vacuum sucks the gas in this chamber out toward the muzzle.

Effectively, the bore evacuator delays the movement of the gas, giving it slightly more time before it follows the path of least resistance. The end result is that the majority of the extra gas goes through the muzzle end instead of back-flowing through the barrel and the effects that come with it. When we asked Gemtech if the bore evacuator reduces the velocity of the ammunition shot, similar to a ported barrel, they told us that if anything it actually increases the FPS. Since the bore evacuator just delays the gas rather than venting it out like an open port in a barrel, this makes a good deal of sense. In order to test it for ourselves we would need another Integra sans bore evacuator, since there are numerous factors that can affect muzzle velocity.

The Dead Foot Arms folding stock makes for a small package while maintaining functionality.

BUILD QUIRKS

But not everything is all daisies. Because the gas block and sleeve are welded into place, you won't be swapping out your handguard system anytime soon. Thankfully, instead of a lower-end handguard they went with a Seekins MCSR V2 handguard. If you somehow ended up with a KeyMod and wanted an M-LOK or vice versa, anything in the Seekins MCSR line can be swapped out. Overall, the modular nature of the AR-15 becomes a skosh less modular with the Integra.

Take note that depending on who makes your M-LOK accessory attachment, you may have to hit the hardware store for some slightly smaller screws to avoid marring the silencer shroud. While everything fits under the handguard, you don't have an awful lot of room on the front half. We had to shave down the screws on the Magpul MVG, and at the end of the day it was easier to remove the rail to add grip covers a la the Troy TRX rail.

Since the upper receiver itself promises to be perennial on the Integra, Gemtech thought it would be best to go with a high-quality known quantity, the Vltor MUR-1A. It's the same internally as any other Mil-spec upper, but rocks thicker walls, is fully heat treated, stress

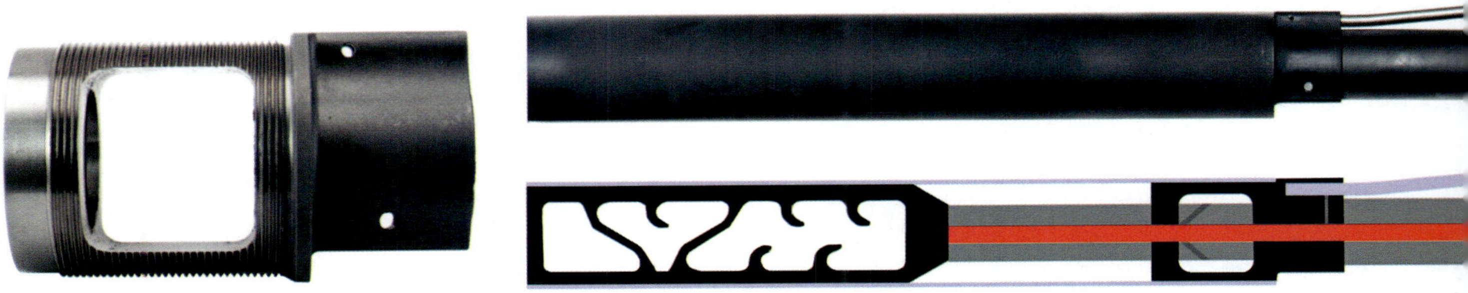

Gemtech's patented bore evacuator.

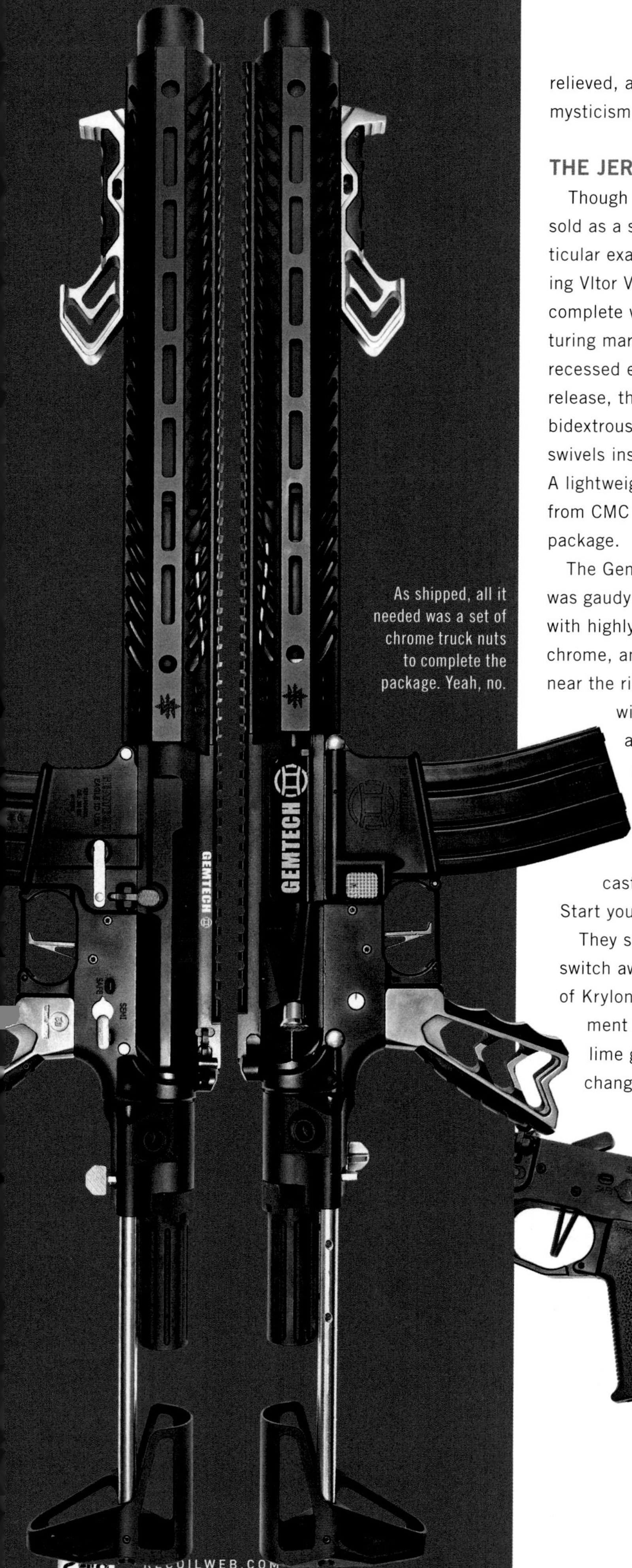

As shipped, all it needed was a set of chrome truck nuts to complete the package. Yeah, no.

relieved, and even subjected to the mysticism of cryogenic freezing.

THE JERSEY SHORE

Though normally the Integra is sold as a separate upper, our particular example has a corresponding Vltor VRA-R lower receiver, complete with Gemtech manufacturing marks. In addition to the recessed extra large magazine release, the VRA has a set of ambidextrous limited-travel QD sling swivels inset into the lower itself. A lightweight two-stage flat trigger from CMC Triggers completes the package.

The Gemtech Integra we received was gaudy to say the least. Heavy with highly contrasted branding, chrome, and polish, this is damn near the rifle equivalent of a bro with a Tap-Out T-shirt and a fistful of energy drink. We wouldn't have been surprised to see it snorting lines with people who could be cast on *The Jersey Shore*. Start your fist pumping now.

They say beauty is just a light switch away or, in our case, a can of Krylon and a box of replacement parts. Like refacing a lime green 1970s kitchen, a change of color can make a world of difference.

We removed and replaced most of the chrome, and swapped out the more egregious branding. Same rifle in a different dress, for sure. But 1,000 times more palatable. Like wearing a Rolex on a NATO strap, there's just something special and fun about using $12 in paint on an expensive rifle.

We won't apologize for focusing on some aesthetics, as rarely does someone cross a street to meet your personality. Since the Dead Foot Arms system adds several inches to the length of pull if used with a standard stock, a shortened receiver extension and buttstock from an LMT PDW stock tightened up that distance. We also liked how the Dead Foot Arms can be fired when the stock is folded, however unlikely it is that particular feature will be needed. In order to achieve this feat, the crew at Dead Foot Arms uses a truncated bolt carrier and a dual spring setup.

AT THE RANGE

It's always nice to know that you're the first one to shoot a rifle. And we know we were because the CMC trigger wasn't resetting. Every other trigger we tried worked fine, while the CMC wouldn't reset with a regular upper either. So, we

dropped in an ELF trigger instead. Once you start playing mucky muck with specialized buffer spring assemblies like you have to with PDW stocks, you can run into issues (for a full breakdown, see our PDW Stock Buyer's Guide in Issue 30). Locking back on an empty magazine was inconsistent before a little spring tuning was performed. Neither were issues related to Gemtech — and you can put the Integra on any 5.56 lower that you want.

Gas to the face? None. Zero. Nadie. Nada. The bore evacuator and tuned gas system are great at their job. On a similar note, usually our internals and magazines get scummed up quickly when shooting suppressed due to the dirty backpressure. We are happy to report that nothing seemed any dirtier than a regular-ass unsuppressed 16-inch AR.

For science, we also tried the Integra without the monocore installed. This isn't recommended because, well, it doesn't work. This rifle is specifically tuned to run with the pressure that the silencer produces. No monocore, no worky.

We tested accuracy with match Prvi Partizan 75-grain ammunition, and the Integra printed a five-shot group at just under 1.5 inches at 100 yards.

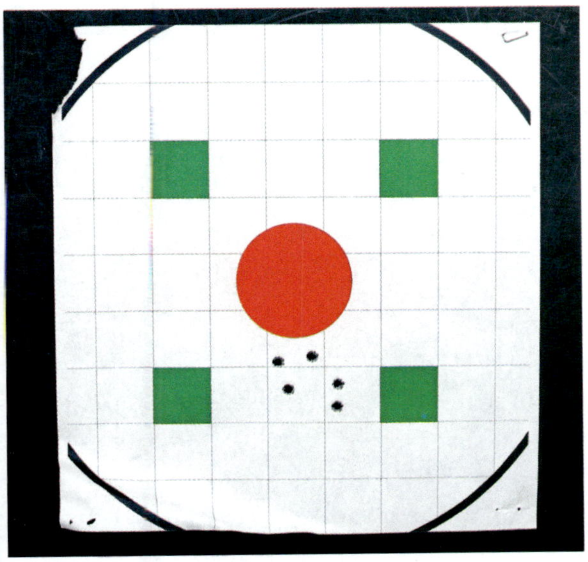

Grouping just under 1.5 inches at 100 yards, accuracy is acceptable but not exceptional.

Regarding sound? We'll rely on folks with $40,000 rigs to make that determination, but Gemtech advertises 131 decibels. This is about on par with an unsuppressed .22LR rifle, and that seems to hold true. We can say that the ejection port noise is greatly reduced, and the Integra was quieter and produced less flash than our other 10.3-inch guns with several different silencers. A nice surprise, since a monocore design isn't widely regarded as being the most effective.

LOOSE ROUNDS

We're not completely sure about the first integral, but arguably the most famous is the Heckler & Koch MP5SD. And the coolest? The OSS' Nazi-snuffing High Standard HDM. As to where the Integra falls onto the list of viable guns — it has a lot going for it. We wish there was more choice regarding handguards. The enclosed front end largely negates the ability to further refine the gas flow, reducing your options to an adjustable bolt carrier gas key like those sold by Rubber City Armory. With that said there's still some flexibility on the rear end of the equation with buffer and spring weights.

When we started, we looked at the Gemtech Integra like a short-barreled rifle with a silencer attached, but it's more than that. The thoughtfulness of design, the fine-tuning, and how it sounds all add up to a package more than just a sum of its parts. Actual. Integral. Hot damn.

A relatively short, quiet, soft-shooting suppressed rifle you can cross state lines with without asking permission from Uncle Sugar? We'll be in our bunk. **R**

GEMTECH INTEGRA

CALIBER: 5.56 NATO

MAGAZINE CAPACITY: 30

BARREL LENGTH: 16 inches (10.5 inches rifled)

OVERALL LENGTH: 26.25 inches (folded), 34.25 inches (open)

WEIGHT: 8 pounds, 6 ounces

MSRP: $1,999 (upper receiver only)

URL: gemtech.com

FEATURED ACCESSORIES:
Vltor Receiver - $380
Elftmann Trigger - $259
Dead Foot Arms Stock - $425
M600DF w/SR07 - $473
Aimpoint Micro - $642
Midwest Industries BUIS - $229
Elevated Tech Mount - $150
VFG & Panels - $56

PRICE AS FEATURED: $4,613

THE EMPIRE CALLED. IT WANTS ITS PDW BACK.

Hokey Religions and Ancient Weapons Are No Match for a Good Blaster at Your Side, Kid

By Iain Harrison

Photography by Henry Z. De Kuyper

Have truer words ever been spoken? Mr. Solo sure knew what he was talking about. Look past the killer custom Cerakote job, and you'll see LWRC's take on what a personal defense weapon (PDW) should be. This is the civilian version of the sub-carbine purchased by the Saudis for internal security purposes. And it rocks.

LWRC
INTERNATIONAL
Six8-PDW
MADE IN THE U.S.A.
SIX 8
EID: 52RX1
PN: SIX8
SN:
PMAG 30SPC

Over the past decades, the submachine gun has fallen out of favor for second line troops, and the PDW's star is ascendant. The requirement to arm cooks and bottle washers with an adequate firearm that is smaller, lighter, and generally less capable than the standard-issue infantry rifle has led to some interesting developments. FN introduced the P90, which has seen limited international sales. H&K's interpretation, the MP7, has found favor with troops that fall outside of its original intended audience. For example, SEAL teams have been employing it in raids on Al-Qaeda franchises in sub-Saharan Africa, due to its reliability, compact size, and controllable nature. And yes, I did just say "MP7." Get over it.

Both firearms utilize a scaled-down cartridge, relying on extended magazines and full-auto capabilities to put enough lead on target. LWRC took a different approach. In conjunction with ATK Federal, who optimized the 6.8 SPC round for use in short barrels, the Maryland company came out with the Six8 IC-PDW, which offers rifle-level ballistics from a very compact package.

Saudi Arabia sits at the crossroads of an extremely volatile neighborhood. While its politics may be decidedly antiquated by Western standards, it remains a key ally in the region, the loss of which would be disastrous for U.S. foreign policy. Following numerous attacks by domestic and foreign terrorist groups, the Saudis decided to get serious about their internal security forces (and drag them kicking and screaming out of their traditional, largely ceremonial role), putting them to serious work. Let's face it, when your issued weapon is a gold-plated MP5K, there's going to be chuckling whenever you walk into the room.

As its name implies, the Six8 chambers 6.8 SPC rounds, which ballistically exceeds that of the 5.56mm.

The replacement for the aforementioned MP5 is the Six8. While not exactly a sawed-off elephant gun, its ballistics exceed that of a 5.56mm by a substantial margin. As evidenced by bruises left on my shoulder from range time, this thing has a decent thump on both ends. And while I wouldn't volunteer to stand down range of any of the PDW cartridges, there's no way in hell I'd go forward of the muzzle of this beast. Launching a 90-grain bonded softpoint at 2,400 feet per second (fps), it's like the Tasmanian Devil of SBRs.

DETAILS, DETAILS...

The LWRC team has managed to pull off an impressive feat of engineering. Not only have they shrunk the AR to very compact dimensions, they also made it run reliably with a bigger round. To achieve this, they ditched the direct impingement operating system in favor of a short-stroke piston housed under the handguards. The usual recoil spring and buffer located in the butt have been redesigned. The buffer is now part of the bolt carrier and, in addition to being reduced in size, contains a second, auxiliary spring and sits inside the rear of the carrier. This has the effect of absorbing some of the force as the assembly impacts the end of the chopped-down

PMAG 30 SPC

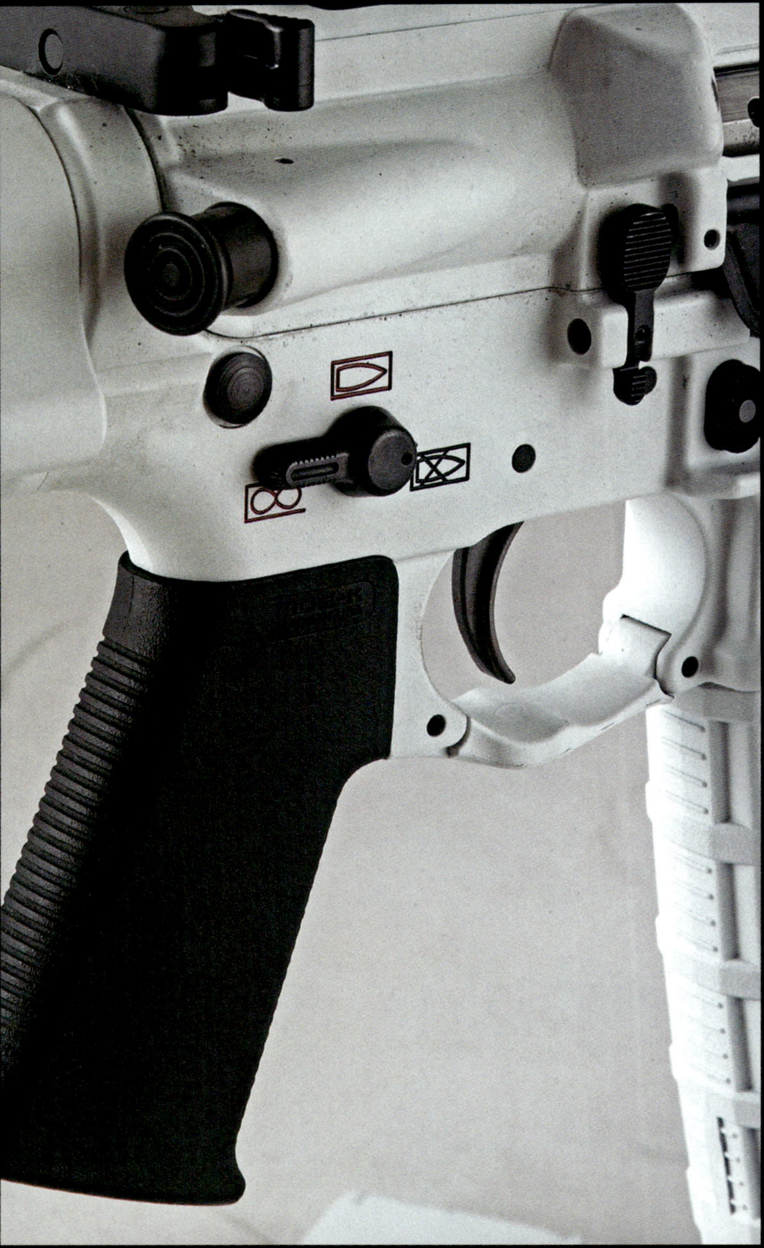

LWRC
PART NO. U07-0294A01
MADE IN U.S.A.

Six8-PDW
MADE IN THE U.S.A.
LWRC
INTERNATIONAL
CAMBRIDGE, MD
U.S.A.
6.8mm SPC

TM
EID: 52RX1
PN: SIX8
SN:

buffer tube. The main recoil spring has been swapped out for one made from flat wire, rather than the round stock typically found in the AR platform, in order to reduce the amount of space it takes up when compressed.

The gas system itself is a modified version of the one found on their full-size guns. It comprises a fixed piston that is part of the gas block, a movable cylinder which slides over it, a spring-loaded operating rod and a short connecting rod which bridges the gap between cylinder and op rod. The gas block's location is dictated by the length of the cylinder and op rod, which no doubt caused some head scratching regarding gas-port diameter, as pushing it this close to the muzzle reduces dwell time considerably. That the gun functions as well as it does is a testament to the amount of development work by LWRC's engineers.

Because of these changes, the field stripping procedure is decidedly non-standard. To tear down the gun, you first clear it and then punch out both takedown pins. In order to separate the upper and lower receivers, you have to pull out the recoil spring which is wound around the rear of the buffer (itself attached to the bolt carrier) as the upper and lower come apart. Although the process is considerably more involved than a regular AR, the good news is that the carrier and bolt stay much cleaner due to this being a piston-operated gun.

Now that the upper and lower are apart, you can turn your attention to the gas system. Unscrew the two knurled and slotted thumbscrews at the front of the handguards. The upper section of the free-float tube can then be slid forward and off, revealing the gas system beneath. Push back on the op rod and remove the connecting link. Slide off the gas cylinder and pull the op rod free of the upper receiver. You can now scrape carbon from the piston exterior and out of the cylinder before reassembling in the reverse order.

The abbreviated receiver extension is achieved via a small buffer that sits inside the bolt carrier, a flat wire action spring, and a second auxiliary spring. The hobbit-sized buttstock telescopes open and closed.

The Six8's lower receiver is hogged out of bar stock and features some neat touches. Fully ambidextrous, the right side has a mirror image bolt hold open, which means that the user can lock open the action without having to shift his grip. This is a major improvement when it comes to dealing with stoppages, as the firing hand can remain on the pistol grip while the support hand racks the charging handle and administers the "tactical tickle." An ambi mag catch and safety complete the controls, while making the gun louder is governed by a proprietary trigger and hammer, both of which are coated in nickel Teflon.

A FRESH MAG

The most failure-prone component of most small arms is the magazine. In designing the Six8, a hard look was taken at the magazines available for it and most of them came up short. While 5.56mm mags can be used in a pinch, the larger diameter of the 6.8 SPC round usually causes feeding problems. Dedicated 6.8 mags are almost always made from steel, as this allows thinner material to be used for the tubes, increasing their internal dimensions. The Saudis, however, looked at Magpul's offerings and said, "We want that." And when you can slap a bunch of petrodollars on the table, people tend to listen.

To accommodate the bigger cartridge, the magazine has been stretched. Due to the thicker walls needed to provide adequate strength when dealing with polymer, they are noticeably larger when you pick one up. As a result, the Six8's magwell no longer accommodates standard USGI mags. While this may be considered a drawback from an interchangeability standpoint, the tradeoff is that the magazine is optimized around the case it's supposed to feed. As LWRC's Joe Devens put it, "We basically designed the gun around the magazine — both lower and upper have been changed to fit the 6.8 round. Magpul's help was a huge benefit to us."

The four-prong flash hider will accept a dedicated suppressor. Note the thumbscrews which secure the upper section of the handguard.

The abbreviated stock slides in tubes machined into a lower receiver extension and is held in either the open or closed positions by a right-side button detent. In order to make the package as compact as possible, the postage stamp-sized butt plate telescopes over the end of the buffer tube. The butt plate itself is nicely machined with grooves on its rear surface and is about as small as it could be and still fulfill a useful purpose.

Folding iron sights adorn the continuous upper Picatinny rail and are a work of art in themselves. Made in-house from 7075 aluminum and type III anodized, they're dehorned and lock in the deployed position. Two sizes of aperture are available on the rear sight, by means of twisting a square post, which is adjustable for windage. Elevation adjustments are handled on the front sight. Another proprietary component is the beefy, ambidextrous charging handle, which features twin, independent latches on either side.

Moving toward the business end, we find a folding vertical foregrip. While not usually a fan of VFGs, on smaller guns they serve the useful role of a hand stop, preventing the shooter's support hand from straying too close to the muzzle. They also keep exposed flesh away from the barrel and gas system as they heat up — this was obviously a consideration for LWRC, as they've fitted insulating panels on either side of the handguard.

ROUNDS DOWNRANGE

It's generally accepted that M855 ammo requires an impact velocity of 2,500 fps in order to fragment — below this velocity floor, the bullet simply drills a 0.22-inch-diameter hole in flesh. An 8-inch barreled AR typically delivers a muzzle velocity of around 2,340 fps, leaving us behind the eight ball from the get-go. This performance can be improved through the selection of a better bullet, but it's illustrative of the problems faced by our troops and the limitations of 5.56 in general. The 6.8, on the other hand, gives us a bigger, heavier bullet, moving at higher velocity. And what's not to like about that? In addition, the Federal round issued with the Saudi weapons uses a bonded softpoint that should give excellent intermediate barrier penetration and terminal performance, far exceeding that of other PDWs.

At the range, the difference between 5.56 and 6.8 SPC was very evident. There's no such thing as a free lunch, however, so the increased smackdown offered by the 6.8 is countered by increased recoil and split times. The usual method to mitigate recoil would be to mount a brake, but with the muzzle this close to your face, you'd regret that decision in short order. Instead, the Six8 is offered with a four-prong flash hider, which also provides a mounting point for a dedicated suppressor currently in development.

If you use a bladed, traditional stance then this little gun probably isn't for you. Even with the stock extended it's too short to use effectively, and in the collapsed position it lends new meaning to the phrase, "nose to charging handle." Tuck it in close to your centerline, however, and it works just fine — though you're not going to

want to burn several hundred rounds in one session, as the butt plate concentrates recoil into a pretty small area. Target transitions are extremely quick and the stock trigger, though quite heavy at about six pounds, breaks cleanly.

While time constraints didn't allow us to put the Six8 through the wringer, we burned enough ammo to come away with a favorable impression. As a means of packing heavyweight ballistics into a tiny space, it's a very competent solution. The Imperial High Command would be proud. **R**

LWRCI IC-PDW

CALIBER: 6.8 SPC

MAGAZINE CAPACITY: 30 rounds

BARREL LENGTH: 8 inches

OVERALL LENGTH: 19.9 inches (Closed) / 24.6 (Open)

WEIGHT: 6.2 pounds

MSRP: $2,300 (Approx.)

URL: www.lwrci.com

THROWING TOMAHAWKS

Honey, I Shrunk the AR ...

By Iain Harrison
Photos by Kenda Lenseigne

The AR's linear piston system is one of the weapon's greatest assets, in that it makes for a smooth, soft-shooting carbine. It's also a liability in confined spaces, as the buffer tube sticking out the back of the receiver means it can neither accept a folding stock, nor compress down tightly to reduce its overall length.

One of the new products that debuted at this year's SHOT Show in Vegas came from Troy Industries, and it manages to shrink the AR to minute proportions. In fact, an NFA-compliant carbine featuring a 14.5-inch barrel with a pinned and welded flash hider winds up shorter than an 11.5-inch barreled SBR equipped with a regular M4 stock, when both are fully collapsed. Which we think is pretty neat.

It manages to achieve this by means of a proprietary bolt carrier, which uses a reduced-diameter tail incorporating the sliding weights and urethane bumper of the usual AR buffer. By stuffing the buffer's guts up the carrier's arse, the entire package is shrunken down, allowing for a shorter receiver extension. Such radical surgery is not without a price, however, as the user is no longer able to hinge open the carbine like a shotgun. Instead, both front and rear takedown pins must be pushed and the upper separated from the lower for maintenance.

The Tomahawk stock itself is reminiscent of HK designs and has four positions, the shortest of which protrudes a scant 3.14 inches from the back of the castle nut. At full extension, it expands to 9.3 inches long, roughly equivalent to the second click on an M4 stock — if you have very long arms, this probably isn't the stock for you, but for the majority, it'll work out just fine. If you choose to use it fully extended, then the lack of a continuous top surface and interruption to a good cheek weld may prove annoying. We tried it at positions 3 and 4, and found it usable in both locations, but the third click was definitely more comfortable.

The Tomahawk is available in three variants, either fitted to complete rifles or as aftermarket parts for you to slap onto your own rifle. The standard $300 version allows for the use of a normal bolt carrier — just unscrew your existing buffer tube and replace it with the Tomahawk and you'll knock 1.75 inches off the previous collapsed length. If you opt for the proprietary carrier, you'll add a Benjamin to your bill, but in return you'll shave almost 3.5 inches. There's a standard version in development for .308 rifles, but at press time had not completed reliability testing.

Our test sample came as part of the Troy Alpha Carbine, which sports a 7-twist, 14.5-inch government profile barrel with carbine-length gas system, as well as the company's Medieval brake, a sling designed by RECOIL contributor Frank

Aimpoint.
TROY

TROY
TROY
TROY
TROY
ALPHA
CARBINE

TROY INDUSTRIES ALPHA CARBINE

CALIBER: 5.56 NATO

MAGAZINE CAPACITY: 30 rounds

BARREL LENGTH: 14.5 inches

OVERALL LENGTH: 27.75 inches (stock collapsed)

WEIGHT: 6.5 pounds

MSRP: $1,300

URL: troydefense.com

FEATURED ACCESSORIES:
Aimpoint PRO - $530

PRICE AS FEATURED: $1,830

Proctor, and a package of Squid Grips, which squish into the rail's perforations and make for a comfortable and insulated gripping surface. The Alpha rail fore-end has its backup front sight machined into the top surface, which makes for a very low-profile setup. While we'd prefer a better trigger than the GI unit the carbine ships with, its $1,300 MSRP probably doesn't leave too much margin to play with.

One thing that concerned us, however, was the use of a carbine-length gas system, coupled with a lighter than usual carrier and buffer combo (12.8 versus 14.7 ounces for a regular BCG and H buffer). At the range, cases were ejected to the 2 o'clock position, which can be indicative of over gassing. We had no failures to feed, fire, or eject during our admittedly short, 200-round shakedown, but we'll continue to keep an eye on it.

For working in and out of vehicles, the Alpha carbine with its Tomahawk stock makes for a pretty sweet setup. It's never going to be as comfortable as a regular stocked gun, but it's as short as you're going to get without resorting to a bullpup. R

LEWIS MACHINE AND TOOL'S COMPRESSOR SBR

Shooters Can Shave Ounces and Inches With LMT's Innovative New Stock System

Story by Iain Harrison

Photos by Straight 8

Once you get talking with the gun geeks at LMT, you start to understand the meaning of the acronym. We toss around the term OCD as shorthand for someone who has a bit more interest in arcane details than we do, but in this case, it's probably not far from its clinical definition. If the first step is acknowledging you have a problem, then we're OK with saying that, yes, we can get excited about the 1907 Roth-Steyr's firing mechanism bearing an uncanny resemblance to that found in a Glock. But these guys take it to a whole 'nother level.

Approached by a NATO member country that wanted a compact carbine for undercover security forces who worked in confined spaces, they examined the problem, then presented a solution that used several of their existing products, as well as coming up with a brand-new one. We got an exclusive opportunity to go hands on with one of the first pre-production units, which coincidentally is one of the guns appearing on this issue's cover.

In developing the Compressor, LMT took their proven L7SA1 upper receiver and stuffed a 10.5-inch, .300 BLK barrel into it. The receiver itself is a monolithic design, with no joint between the upper and handguard, making for a very strong unit with almost no chance of force being transmitted to the barrel nut from the rail system. It's what the AR would have been if Eugene Stoner had access to five-axis CNC machines.

The .300 Blackout was chosen due to its versatility and ability to be effectively suppressed, an essential requirement of the end user. Due to their environment, if they'd lit off, say, an unsuppressed short barreled 5.56, communication between team members would have been compromised. Most of them would have been deaf after the first few rounds.

In addition to several advantages, the .300 BLK brings with it a few problems. We've messed around with enough examples over the past half-decade to be able to look past the hype and sycophantic fanboy-ism and realize that while it's a decent compromise that enables the AR-15 to launch heavy, subsonic bullets, it's not a magic death ray. When setting up an upper, it's usually necessary to decide whether its primary ammo is going to be super or subsonic, as either longevity or reliability is going to be compromised if the wrong choice is made. Subsonic ammo is usually more reliable when teamed with a pistol length gas system, but firing supersonic ammo in the same setup can result in an upper which is over gassed — problems such as the bolt outrunning the magazine springs can rear their head.

Conversely, an upper using a carbine length gas system (which depending on barrel length is optimum for supersonic ammo) can be very picky about supersonic fodder and may produce stovepipes when fed ammo it doesn't like. Add to this mixture variations in back pressure when a suppressor is mounted or removed, and you have an engineering nightmare.

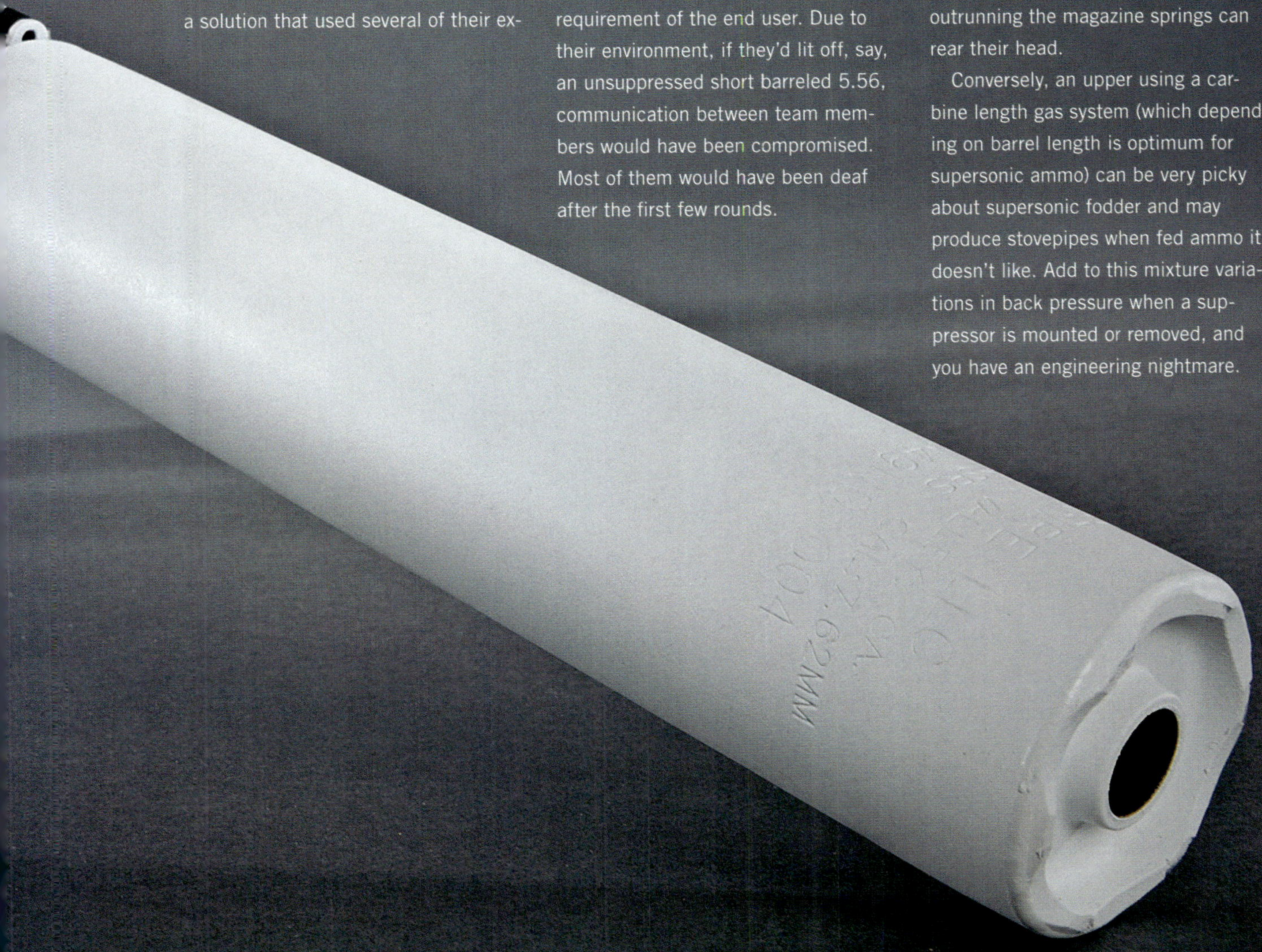

The Blackout is typically fueled by a slow pistol powder, which by rifle standards burns much faster than usual. Because pressure spikes closer to the breech than with the original 5.56 round, by the time a bullet passes the usual gas port location, it's running out of steam. The acid test for any .300 BLK gun is its ability to work with subsonic loads when unsuppressed, and supersonic loads when fitted with a can. These combinations produce the widest range of gas pressures and volumes, presenting a tough hurdle for any manufacturer to overcome.

The Internet would have you believe that a 9-inch barrel is optimum for this caliber, but in our experience this is not the case. You can gain a bit more performance with a longer tube without producing an ungainly weapon. LMT seems to have reached the same conclusion, as the Compressor uses a 10.5-inch barrel, allowing the gas system to remain pressurized for a longer period of time than the industry-standard option. Offsetting the slightly longer barrel is LMT's new stock and buffer system, which chops 3 inches off the overall length of the gun.

Despite its flaws, the AR-15 endures as the preeminent American rifle of the 20th and, so far, 21st centuries. *RECOIL* readers are no doubt familiar with the limitations of the multi-lug bolt, the craps-where-it-eats gas system, goofy charging handle, and tits-on-a-fish forward assist. Perhaps not as well known, but more important from a functionality standpoint, is the action's short dwell time.

Go ahead and fully retract the BCG on your AR. You'll no doubt notice that compared to say an AK, there isn't that much distance between the back of the magazine and front of the bolt. When it comes to reciprocating components, distance equals time, and that ¼ inch or so just doesn't give the magazine much of it — in order to push another cartridge into position before the bolt runs forward

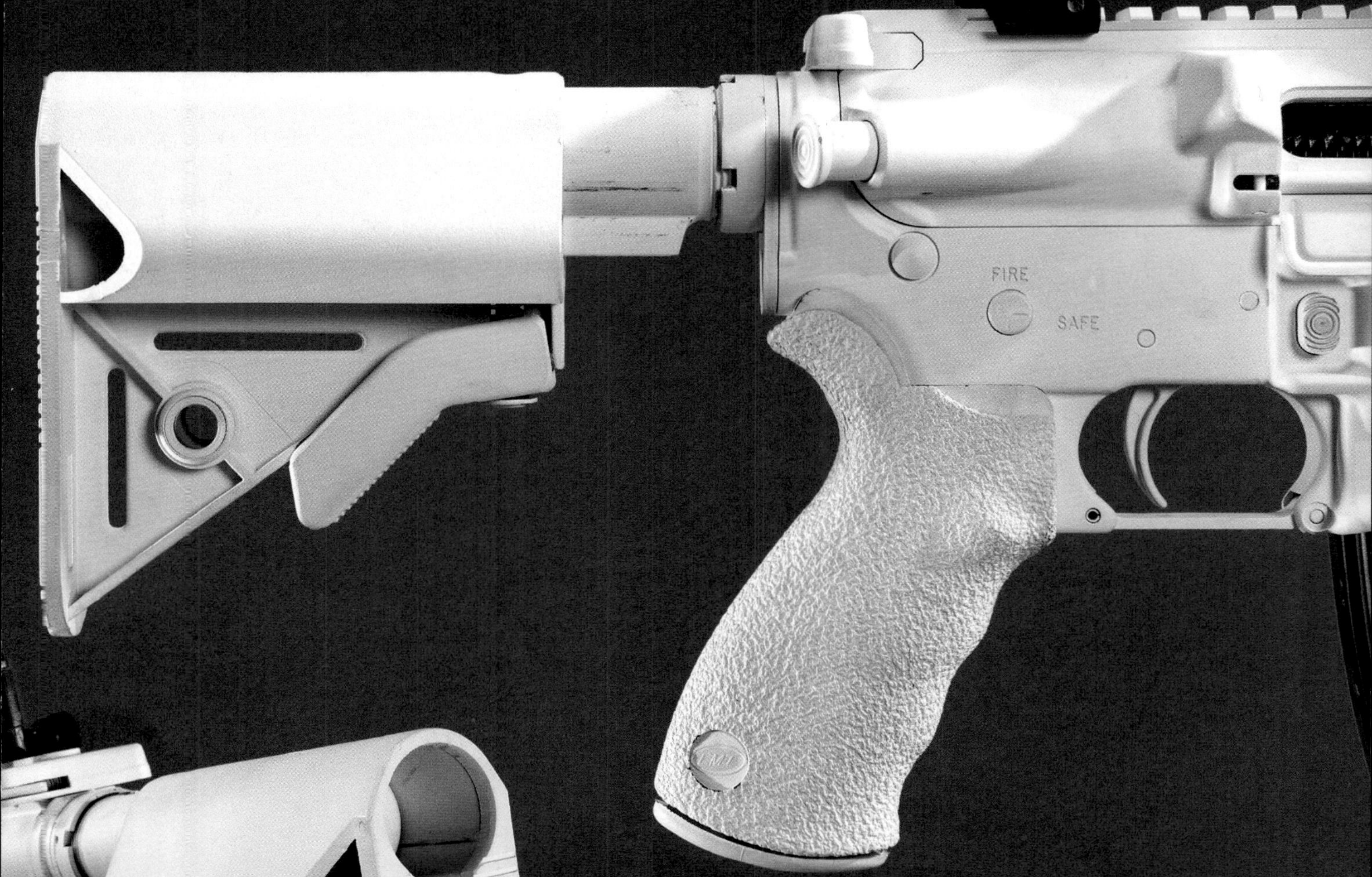

after reaching the end of its stroke. We can slightly change how quickly the BCG returns by adding or subtracting mass.

One of the ways manufacturers have sought to reduce the incidence of bolt over base stoppages has been through increasing the number of sliding weights in the buffer, thereby reducing its tendency to bounce off of the end of the receiver extension. Any change to the AR's design that fails to account for the action's short dwell time will negatively impact reliability.

Which brings us to the LMT Compressor's buffer system. Their engineers managed to shorten up the buffer tube without shortening the BCG's stroke, by means of chopping down the buffer's length — and they did this without losing any reciprocating mass or the deadblow effect of the original's sliding weights. Which we think is pretty trick.

When you break the system down, you quickly reach a forehead-slapping moment and question why you didn't think of it first. Immediately apparent is the abbreviated SOPMOD buttstock, which slides onto a shrunken, six-position buffer tube. Despite losing inches, there's no loss of functionality. What's hidden from the outside is the patented bit of the system, consisting of a tiny aluminum buffer, a flat-wound recoil spring, and a bolt carrier insert. It's this insert that slides into the open rear end of a regular bolt carrier and houses a sliding tungsten weight. The beauty of the system is that it requires no permanent changes whatsoever to a stock AR-15 upper, and can be retrofitted to any existing carbine. Like we said, forehead-slapping.

ROUNDS DOWNRANGE

Lewis Machine and Tool will be introducing their own suppressor design in 2016, but it wasn't ready for our range session. So we enlisted the help of SureFire, who sent along one of their SOCOM300 SPS cans. After slipping it over the matching muzzle brake, we slapped in a magazine of 220-grain subs and started hammering steel. So far, so good. Removing the suppressor and remembering to insert earplugs, we then shot it unsuppressed and noticed that brass was ejecting into the same pile that the first mag produced, indicative of both a low back pressure suppressor and a well tuned gas system.

The third magazine was loaded with 120-grain supersonic rounds, and we got through about half of it before the RECOIL curse manifested itself. We're well known for breaking shit (or maybe it's just that when stuff breaks, we actually tell you about it), and true to form, the prototype carrier insert decided to spontaneously disassemble itself, locking up the bolt. LMT's representative was quick to jump on a call with their engineering department, and we discovered that several changes had been made to the production version.

In ours, the tungsten weight was retained by a simple threaded cap that had backed out, whereas the cap on newer ones are also secured by high-temperature thread locker and a couple of spot welds. Despite lacking the new improvements, our proto had still managed to last through several range sessions prior to its arrival in Arizona before letting go, leading us to believe the new version should hold up. But as always, we'd like to test a production unit before pronouncing it ready for service.

We still had ammo to burn, so rather than call it a day we installed the upper on a different lower and shot it with both supersonic and subsonic rounds. Because no one wants to leave a range with anything other than brass in their range bag. Apart from a slight problem caused by our reverse Midas touch, the session went as planned and no further problems were encountered.

We're always conscious that being the first to try out new gear brings with a certain amount of risk, for both writer and manufacturer alike. We think the trade-off is worth it, as you get to see cool kit before anyone else and see potential pitfalls and what's been done to correct them. In this case, LMT's stock system is a novel and innovative approach to

solve an end user's problem. The Compressor stock will be available as an aftermarket package that can be fitted to any AR-15, and should bring down the collapsed length of a 14-inch, pinned and welded carbine to 26.5 inches — just above the legal minimum, and equal to many SBRs. What would you rather have: 3 inches of additional barrel or 3 inches of buttstock? Thought so. ■

LEWIS MACHINE AND TOOL LM8CSW300 COMPRESSOR

CALIBER: .300 Blackout

MAGAZINE CAPACITY: 30 rounds

BARREL LENGTH: 10.5 inches

OVERALL LENGTH: 24.3 inches (collapsed)

WEIGHT: 7.3 pounds

MSRP: $1,300

URL: www.lmtstore.com

FEATURED ACCESSORIES:
Shield Sights LTD CQS 2MOA - $370
SureFire SOCOM300 SPS - $1,075
Conversion Kit - $425

PRICE AS FEATURED: $4,561

SHOULDER-FIRED FLASH BANG

POF's 7.62 NATO MK18-Sized Revolution is for Great Lighting Barbecues and Putting 30-Caliber Holes in Stuff

By Karl Kasarda

Photos By Kenda Lenseigne and Karl Kasarda

It's interesting to observe what garners attention within firearms development … and what doesn't. Flashback to 1957 — the design of the rifle hadn't changed much for a century. Certainly, technology has advanced, with the percussion ignition-rifled musket leading to self-contained cartridge-fed repeating actions and ultimately semi-automatic battle rifles. But the rifle throughout all of the evolution was still recognizable as a rifle: wood and steel.

There were many key developments in World War II; the advent of stamped sheetmetal receivers and the intermediate cartridge were two of the biggest. Those evolutionary steps were the desperate products of two authoritarian regimes across the pond attempting to outproduce one another, with easily mass-produced weapons not just in flesh in blood. But Americans still had a semi-automatic battle rifle, a wood and steel creation that fired the full-sized .30-06 cartridge. We had the resources and capability to produce the M1 in large numbers without modern, mass production methods. Nor was there much impetus to consider an intermediate cartridge; the USA just defeated Imperial Japan and fascists at the same time with that combination. It was the "greatest battle implement ever devised," per General Patton himself.

It's hard to swim against such a strong tide, but that's exactly what Eugene Stoner and Armalite did in 1957 with the AR-10. The use of modern materials, such as aluminum and fiberglass, along with aerospace engineering methodologies, resulted in a rifle that no longer looked nor felt like what one expected from a rifle. Skepticism and bias from the military hierarchy was immediate. One of those futuristic experimental ideas, a composite aluminum and steel barrel to reduce weight, was a bit too much so and ruptured during the trials, giving the military an excuse to promptly discard the AR-10 as nothing more than a gee-whiz prototype. Thus ended the possibility of its adoption. Back to the old wood and steel we went, and the old-school thinkers were placated.

In summary, the AR-10 was too different, too unusual. It deviated from accepted norms too much to be tested without bias.

Fast-forward to today, and the market is filled with endless AR-10 and AR-15 "variants" that are essentially identical. Certainly quality control, features, and aesthetics vary, but under the hood, they're mostly the same. They've become the equivalent of rifle cordwood. It's become difficult to determine what's different, if anything, between one AR manufacturer and another. Ironically, the AR's forward-leaning design concepts from 1957 are now the baseline of what people equate with a rifle — a truly 180-degree culture shift from then to today.

This prevents some very valid attempts at evolving the AR platform from breaking through the clutter; if it looks like an AR, it's just another AR. But what if that's not always the case?

Looks like a 5.56mm, but certainly ain't.

THE REVOLUTION

Patriot Ordnance Factory has made some very interesting evolutionary changes with the AR platform, both the 15 and 10, that go relatively unnoticed unless you pay close attention. POF started initially with piston guns, but in this article we focus on their direct impingement rifle, the Revolution.

POF's 308 NATO version of the Revolution is dimensionally identical to the AR-15 5.56 equivalent (excluding the magazine well) and still uses an AR-15-sized bolt carrier group — for a gun firing 308. This results in a smaller, lighter 308 rifle that weighs right around 6 pounds. This is what Armalite attempted to bring to the table with their original AR-10, but ultimately failed to achieve until Jim Sullivan and crew downsized the rifle to the 5.56 AR-15. So with the Revolution's weight and size, POF accomplished what Armalite intended to do in 1957.

The weight of the 308 DI Revolution, as well as the dimensions of this gun, isn't immediately noticeable until you handle it. It's remarkable that it has the heft of an early AR-15, while being chambered in the round that the AR-10 was intended for. The 308 DI 16-inch Revolution comes in at 6.81 pounds, while we got to test a prototype 10.5-inch SBR, a mere 5.9 pounds.

POF has implemented other interesting modifications as well. Their E2 dual-extraction technology applies the fluted cylinder concept, most commonly associated with HK and delayed roller-action guns, to the neck of the cartridge chamber.

This fluting allows for gas to break the seal formed around the neck of the cartridge, aiding in extraction during high-pressure situations. While generally not as significant an issue with 7.62x51, the relatively straight-walled 5.56 cartridge is well known for being a bit challenging to extract, especially when things aren't quite right. This can result in a stuck case in the chamber, rendering the rifle temporarily unusable. We'd like to put POF's E2 idea through more testing, but fluted chambers are proven to work, and this implementation around the cartridge neck makes logical sense.

Another significant, but also easy to miss, difference with the POF design is their heat sink barrel nut. Their barrel nut is a full 3 inches in length, providing a substantial amount of surface area for cooling. In a DI gas system AR, the heat of the gases is transferred down the gas tube into the gas key, bolt carrier group, and bolt itself. The bolt is, of course, locked into the barrel recess during firing, and the cartridge itself transfers heat to the chamber during the firing cycle. The enlarged barrel nut is in close proximity with all of those elements of the gun and acts as a heat sink; adding length and surface area to that component is a very slick way to transfer heat away from the action of the DI gun. This should add to reliability and reduce heat-induced wear and tear on those components.

POF included a number of other improvements, such as a roller cam pin to reduce friction and wear, a monolithic bolt carrier group to eliminate the possibility of a loose gas key, easily adjustable gas system, and a myriad of coatings across different components within the system to reduce wear, increase hardness, and so forth. A debate about the different coatings alone could fill a book. The point is there's more going on inside the Revolution than meets the eye; we've just highlighted some of the more significant elements.

No irons here; we outfitted this POF Revolution with an optic only.

ROUNDS DOWNRANGE

Unfortunately, I didn't have an extensive amount of time with the

gun on the range — such is the nature of prototypes — but did shoot a number of magazine dumps and rapid fire within quick succession to assess the handling of a 5.9-pound 308 NATO AR carbine.

The first shot fired was a little jarring. The short muzzle brake coupled with a significant amount of unburned powder in such a short barrel resulted in a lot of that 308 power manifesting itself as flash and boom just in front of the shooter. It only took a moment to adjust to it, but I'd recommend something other than a brake for any SBR 308; muzzle devices like the four-piece Bulgarian AK74SU flash hider come to mind as a possible solutions.

Once you become accustomed to the muzzle's pomp and circumstance, you can really appreciate the 10.5-inch Revolution's attributes. Let's be clear, though — 308 is still 308, and anyone who states, "This 308 recoils like a 5.56," is being hyperbolic. However, the recoil attributes of the little POF are startlingly mild for its weight and size. The design attributes as manifested by Stoner himself still apply, but removing a lot of the reciprocating mass of what would normally be a 308 bolt and carrier and replacing it with 5.56 equivalent components means less to deal with in the recoil cycle. Performing a full 20-round mag dump on a steel mini-IPSC at 50 yards was an easy task. Double taps were also quickly and easily dispatched.

Imagine a gun that feels like a 5.56 MK18, but fires full power 7.62x51. The Revolution was controllable, light, and handy. It has to be experienced to really appreciate; it's not obvious by appearance alone.

This rifle also doubles as a firestarter.

Though the carriers are the same size, the round isn't.

We didn't have enough time with the gun to address durability and reliability. However, no issues occurred during our firing session — and we used only the highest quality Wolf steel-cased 308. While I don't advocate Wolf as go-to ammunition for serious purposes, it certainly does reduce the cost of training and practice. It's good to see if a rifle can run this type of ammo reliably, and the POF 10.5-inch SBR did so flawlessly.

It's worth noting that standard 308 NATO ball is a bit much in a 10.5-inch barrel. At this abbreviated length, standard 308 NATO pushes a 147-grain projectile at just under 2,300 fps, about what you'd expect from a load developed for 20-inch barrels. This approaches 7.62x39 in performance, so that's worth consideration. But it's also worth thinking about how ammunition could evolve to cater to such a platform, now that it exists. Imagine what you could achieve in a rifle that's the size and weight of a 5.56 carbine if ammunition and powder manufacturers collaborated to squeeze all the potential out of it.

IMAGINE A GUN THAT FEELS LIKE A 5.56 MK18, **BUT FIRES FULL POWER 7.62X51.**

Imagine the Leverevolution or CFE BLK mentality applied to 7.62x51. Hornady, are you listening?

At the time of writing, POF is still considering putting out a 10.5-inch variant of their DI Revolution. If they do, it'll be available as an SBR as well as a pistol. In the meantime, all of their evolutionary modifications and ideas are still available, albeit in a slightly larger form. But should you like the idea of a carbine that can also double as a flamethrower, then please let them know. The beauty of a free market economy is that if you have both desire and cash, someone will step up to satisfy the former while collecting the latter. **R**

ABOUT THE AUTHOR

Karl Kasarda is a long-time training junkie and competitive shooter in multiple disciplines, as well as the progenitor of the Two Gun Action Challenge match (2G-ACM). He's also the match director and creator of Desert Brutality and Finnish Brutality, both of which are major two-gun events in their second year running. Karl is the host and owner of InRangeTV, a history and firearms culture-related social media network. When he isn't shooting, running a match, or creating videos, he's hacking or researching something historical or esoteric.

PATRIOT ORDNANCE FACTORY REVOLUTION DI

CALIBER: 7.62 NATO

MAGAZINE CAPACITY: 20 rounds

BARREL LENGTH: 10.5 inches

OVERALL LENGTH: 26.75 inches (stock collapsed)

WEIGHT: 5.9 pounds

MSRP: $2,700

URL: pof-usa.com

FEATURED ACCESSORIES:
SIG Romeo4S red dot - $540

PRICE AS FEATURED: $3,240

HEY, CU

The Ares Defense SCR Takes AR Mags and AR Uppers — and is New York Legal

By Peter Palma

Photography by Kenda Lenseigne

If you happen to live in a state like New York, California, or Connecticut (where most of the stuff in this publication is illegal for you to purchase), you still have a few options left. Assuming moving to a free state isn't one of them, there are nonetheless semi-automatic 5.56mm rifles available to you. And Ares Defense has recently added one more.

Historically, the Ruger Mini-14 has always been available to the entire United States. But, with its proprietary, expensive, and difficult-to-find magazines, questionable accuracy, limited aftermarket support, and less-than-ideal optic mounting, it hasn't been an optimal choice. The Kel-Tec SU-16 series rifles at least accept STANAG magazines and feature Picatinny rails for mounting optics, but the shape of the stock requires them to be mounted lower than an AR, so a good cheek weld with an ACOG or optic of similar height is impossible. Also, the Kel-Tec has even less aftermarket support than the Mini-14. What about the venerable AK? Well, those neutered Saigas in 5.56mm caliber also have proprietary magazines, little aftermarket support, and inadequate mounting options for optics.

So the products available to people in these benighted states just don't give the shooter too many choices. And as American consumers, we value choices, damn it! We want at least three gas stations on every street corner, 200 craft beers on tap, and 1,000 toothbrush options to choose from.

There are a number of modifications available for the AR-15 platform that are legal, but they tend not to be very ergonomic, as they're an adaptation to bypass — sorry, *comply with* — ridiculous legislation. They're also ugly. Which is kind of irrelevant for a firearm but, hey, you still have to look at it. Well, now citizens behind enemy lines can rejoice as there is a solution in the Ares Defense SCR, or Sporting Configurable Rifle. The SCR is a 50-state-legal, lightweight carbine that accepts STANAG magazines and AR-15 uppers.

JOMO...

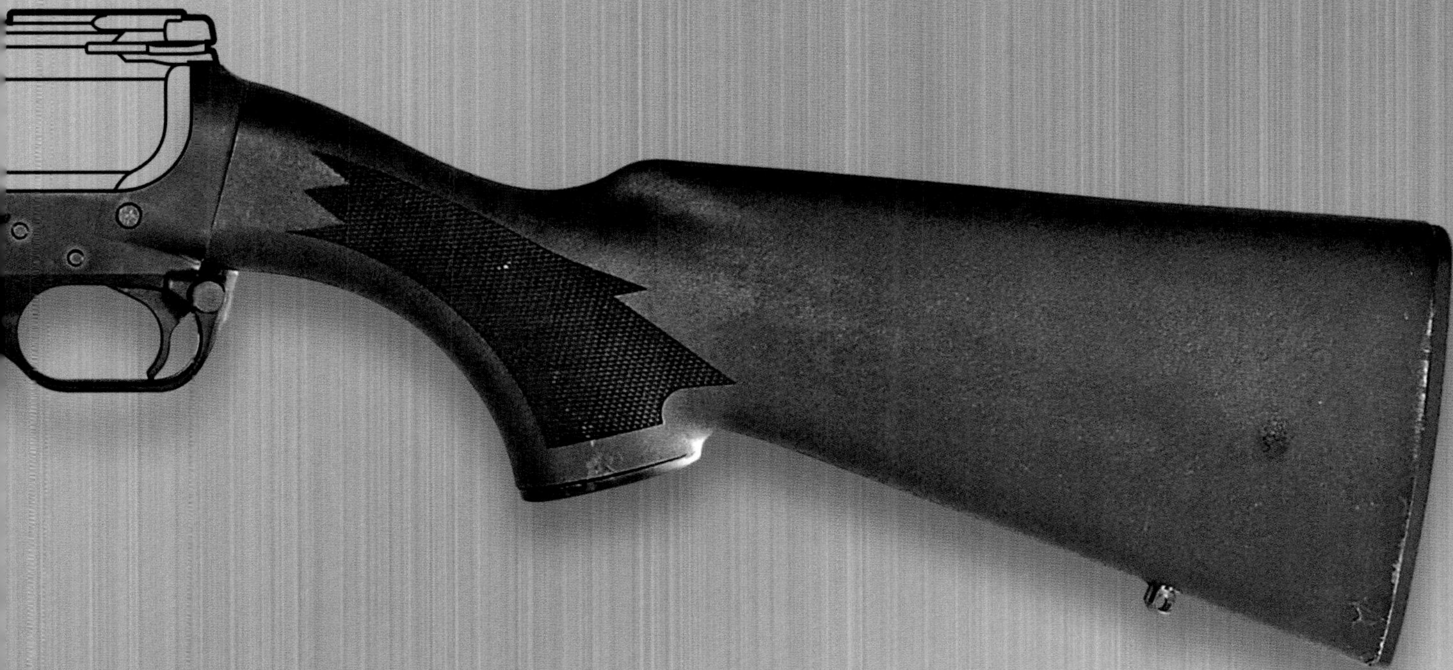

THE SCOOP

If you're looking at the photo on this page and thinking "impossible," let us explain how the engineers at Ares Defense figured this one out. The lower borrows the hand placement and stock style of a hunting shotgun or rifle and marries it with the magazine well and takedown pin locations of an AR-15. To make the grip as comfortable as a traditional firearm, the recoil spring is located in the stock and goes down at an angle through the pistol grip. The receiver is designed to interface with Remington 870/1100 stocks, but if you want to add your own favorite piece of highly figured walnut, be warned that you may have to remove a tiny amount of material from it — but apart from a minor bit of fitting, it should be compatible. The proprietary bolt carrier has a hinged tail similar to a FN FAL and travels in the direction of the aforementioned recoil spring.

The SCR is only sold as an entire rifle, rather than a stripped lower, and the base model is equipped with a non-threaded, 16.25-inch, lightweight contour, 4140 Chromoly 1/9-twist barrel, along with a five-round magazine, Magpul MOE handguards, and a Monte Carlo stock that will enable a decent cheek weld — even with taller optics. The SCR is also available with a sporter stock (with lower comb) and a short sporter stock for smaller-framed shooters. In the interest of science, we found that a powerful buttstroke could more easily be delivered from the SCR than an AR-15, as the shooter did not have to adjust his grip.

Since the popular Remington 870/1100 stocks fit on the SCR, the shooter has a wide variety of styles and colors to choose from. A wood buttstock could be combined with a wooden AR-15 handguard (yes, they are available) to make a very non-threatening firearm that still has modern capabilities whilst keeping a low profile. The SCR is also available in 7.62x39mm and can be fed from C Products Defense

The SCR's muzzle is not threaded to maintain compliance with laws such as the New York SAFE Act. Note the recessed ring, which will accept clamp-on muzzle devices.

stainless-steel magazines, which in this writer's experience are arguably the only AR mags in this caliber worth a damn.

The standard SCR is not equipped with a last-round bolt hold-open; however, this can be purchased from the factory and fitted as an option. The magazine release is more difficult to reach than on an AR-15, making emergency reloads slower. The situation is reversed in the case of tactical reloads where the shooter retains a partially empty mag, once the optional ambidextrous magazine release is added — you hit the mag release with your support hand thumb while simultaneously stripping the partial mag out of the gun

Because of the way the rear of the lower is shaped, popular .22LR conversion kits or uppers cannot be used with the SCR rifle without a good deal of modification — or much bravery with an angle grinder. To address this shortcoming, Ares Defense is developing a compatible .22 upper, but there is no word yet as to when the release date for this might be. Perhaps around the time affordable .22LR ammo shows up...

With the use of a sporter stock, the shooter can utilize very low-mounted optics and take advantage of the slightly extended point-blank range afforded to firearms that have

a very small optical (offset) height. Our trigger-pull gauge maxes out at 12 pounds, and the SCR's trigger was literally off the scale. We felt it probably came in at around 15 pounds, which was a bit much for precision work. The reset was audible and smooth, but not very forceful, still leaving the shooter with more than a couple of pounds to take up before the break. The trigger functions basically the same way as semi-automatic hunting rifle or shotgun, so a properly trained gunsmith should be able to slick it up. But for those of us used to the simplicity of a drop-in solution, the good news is that Bill Geissele is working with Ares Defense to produce an aftermarket version. The SCR has a cross-bolt safety reminiscent of a Remington 870 — easy enough to disengage with the trigger finger, but not as easily reapplied as with an AR. Any charging handle can be utilized with the SCR, though it comes from the factory with a perfectly adequate ambidextrous version. The shooter should be cognizant of not choking up on the pistol grip too far, as this can obstruct the charging handle's rearward travel. This is one of those self-correcting problems, as after the first couple of gashes to the thumb, you tend to remember to shift your grip while racking.

The lightweight contour barrel played a large part in keeping the test rifle's weight well below the 6-pound mark. Even with the addition of an Aimpoint Micro T-1, it tipped the scales at a featherweight 5.8 pounds. However, the low-profile gas block doesn't have provisions for mounting a front sight, so backup irons are out of the question. The more observant among you will no doubt have picked up on the lack of a flash suppressor and the barrel's recessed ring, machined near the muzzle. Because threaded muzzles

The sporter stock works perfectly with low-mounted optics, such as the Aimpoint T-1 pictured here, resulting in a small optical offset.

Above: AR-15s have a receiver extension into which the bolt carrier group, buffer spring, and buffer reciprocates. The SCR instead has a proprietary bolt carrier with a rat tail that extends into an angled recoil tube and spring in the stock when it travels rearward, similar to the FN FAL. The SCR bolt is shown on the left, and an FN FAL bolt on the right. Note that this example lacks the optional bolt hold open.

are one of the "evil features" banned by New York's SAFE Act, an alternative solution was needed. "No problem," said the engineers at Ares. "We'll just design a clamp-on flash hider that doesn't need muzzle threads."

The upper thankfully comes without a forward assist and is a slick-sided unit bereft of a brass deflector, though aftermarket brass deflectors that attach to a Picatinny rail are available. We realize that we are nitpicking, and in reality none of these small issues with the upper really matter because a shooter can — and probably will — simply pop the takedown pins and drop on any AR-15 upper they like. Note, however, that the bolt carrier for the SCR is proprietary, so you cannot use standard or aftermarket AR bolt carriers.

Top: While only sold as a complete rifle, the SCR's lower will accept any AR-15 compatible upper, such as the homebuilt one shown here. You will need to swap out the bolt carrier, though. Be sure that your choice of components on your replacement upper complies with local laws. The Monte Carlo stock with raised comb accommodates typical tall AR optics.

RINGING STEEL

Out on the range the SCR, combined with the Aimpoint Micro T-1, delivered consistent 3-inch groups at 100 yards with both Winchester Q1311 and American Eagle ball ammo, with our largest group coming in at 3.25 inches. MGM auto poppers were boringly easy to hit at that range, and the light weight of the rifle made it easy to transition between targets and to shoot from the standing position. There was no additional perceived recoil from the unorthodox bolt and recoil spring setup, and the overall performance was on par or above any semi-automatic 5.56mm rifle with the same $850 MSRP. Shooting an SCR in comparison to an AR-15 is relatively similar and unremarkable, which is the whole idea behind its design.

We wanted to check out the SCR's adaptability, so we grabbed a homebuilt upper with a FN M-4 barrel, rifle-length handguard, and a higher mounted Aimpoint Micro T-1. Pulling the pins on an SCR lower fitted with the Monte Carlo stock, ambidextrous magazine release, and bolt hold-open option, we ditched the factory upper and slapped on the new one. With observers standing by, we waited for the Dr. Jekyll-to-Mr. Hyde transformation to overtake our psyche, for an inanimate object to fill us with blood lust. Nope, nuthin'.

Running this setup was even more seamless than the stock SCR. Magazine changes were much faster and safer, as we were able to close the bolt with the release button, rather than using the charging handle. Our five-shot group at 100 yards even shrunk to 2 inches, which isn't

bad considering we were using ball ammo, an unmagnified optic, and a trigger the weight of a bowling ball. No stoppages of any kind were noted during the range session.

BOTTOM LINE

The Ares Defense SCR is a viable and ergonomic rifle that gives the owner access to a vast aftermarket of magazines, parts, and accessories. Until now there has been no rifle designed specifically as an alternative to the AR-15 platform while offering compatibility with many of its components.

With certain pipsqueak fascists pushing legislation that prevents upright citizens from owning the types of rifles needed to ensure the security of a free state, responsibility is falling on manufacturers to create alternate products. We're sure it isn't lost on any law-abiding individual that the Ares Defense SCR complies scrupulously with the law in states that have passed draconian restrictions. Or that AR-15 uppers are not in themselves firearms, and therefore freely available. R

The SCR performed very well at the range, feeling very similar to a typical AR-15. Which is exactly the point.

ARES DEFENSE SPORT CONFIGURABLE RIFLE

CALIBER: 5.56mm

MAGAZINE CAPACITY: 5 rounds

BARREL LENGTH: 16 inches

OVERALL LENGTH: 37 inches

WEIGHT: 5.7 pounds

MSRP: $850

URL: www.aresdefense.com

FEATURED ACCESSORIES:
Aimpoint Micro T-1 - $753

PRICE AS FEATURED: $1,603

FROM NEW HAMPSHIRE WITH LOVE?

SIG's MCX SBR Turns 7.62 x 39 Red and We Love It

By Rob Curtis

While the term Virtus dates at least 600 years before the birth of Christ as a way of describing the virtues of Rome's fighting men, it feels like it took the 7.62 x 39mm version of SIG's namesake MCX Virtus nearly as long to reach the market.

But, it's arrived. And, like Horatius at the bridge, the MCX x39 earns the heroic Virtus commendation by standing tall amid an onslaught of thousands ... of rounds ... of mostly garbage ammo, that is.

LOOKING BACK

This isn't SIG's first rodeo building a modular platform with the ubiquitous Russian cartridge in mind. It's hard to ignore the company's redheaded stepchild, the SIG556xi and its stillborn AK magazine compatible variant. Yet, leave out the misguided attempt to capture the nobility of SIG's SG 550 series rifles in a U.S.-made package and concentrate on the configurable nature of the rifle, and you might see parallels in the company's vision for the MCX platform.

While the SIG556 was midway through

its product life cycle, evolving into the configurable 556xi, what was to become the MCX began development in response to a government contract looking for a low visibility assault weapon (LVAW) chambered in 300 Blackout. The two products were developed in separate silos. The former was a commercial product line extension; the latter, a product meant only for the military market. Yet, both grew into configurable, multicaliber, short-stroke piston platforms that would each sprout a viable 7.62 x 39 offshoot.

As the LVAW matured into the MCX platform, the Highlander principle played out and the company shifted development and production resources to the more popular platform. The 556xi, and its nascent 7.62 x 39 Russian configuration had to die in order for the MCX to fulfil its destiny.

WHY 7.62 X 39 NOW?

There are few reasons for 7.62 x 39's popularity. One is its worldwide ubiquity. Another is its performance against vehicles.

SIG's SRD762-QG silencer is remarkably effective at taming the MCX SBR x39's stout recoil.

Like all ammo in the past decade, x39 ammo prices have been on a rollercoaster. SIG's putting out the MCX Russian as the price of x39 in the U.S. seems held at a tolerable, if not historically agreeable, level. So the timing of the domestic release makes economic sense.

But, we're guessing foreign military sales is where the MCX x39 is going to shine. There's gotta be mountains of 7.62 x 39 in onetime Soviet allied nations. Some of those nations are joining NATO and their military forces are likely interested in a rifle platform that allows the use of both stockpiled 7.62 x 39 ammo and newly acquired 5.56x45 NATO with only an armorer-level parts swap needed for compatibility.

The economic benefits of running the x39 round are pretty straightforward, but there's also a specific performance dimension to consider. Military and law enforcement types working around cars will tell you x39 is what you want. This point is affirmed by Doug Flavin, a SIG Academy senior instructor and the Academy's armorer program manager who says, "You're getting far greater through and through vehicle penetration with 7.62 x 39 than you are with 5.56mm. That's unequivocal."

ONLY IN MCX X39

Digging into the 7.62 x 39 SBR version of SIG's modular carbine, there are only a handful of differences between it and its western chambered brethren. Mike Joslin, SIG's assistant rifle product manager, explained what's going on in the guts of the MCX Russian.

The MCX Russian is purposefully designed so that nearly all of its parts are interchangeable with any generation of 5.56 NATO and 300 AAC Blackout versions of the MCX. That being said, Joslin points out the parts that are unique to the x39: the operating rod assembly, bolt, barrel, and magazine.

In order to maintain the reliability of the MCX with typically stubborn Eastern European primers, the x39 version uses a heavier hammer spring and lacks the MCX Virtus' firing pin lock. The latter is removed from the op rod assembly to maximize the benefit of the former since activating the firing pin block takes energy from the hammer. "We needed every bit [of energy] we could to get reliable ignition across the vast differences in primer

sensitivity seen across 7.62 x 39 ammunition."

Because the x39's op rod doesn't have the firing pin lock mechanism present on the 300 AAC and 5.56 NATO versions of the MCX, SIG changed its length to prevent the carriage assembly without a firing pin lock from running in the wrong caliber rifle. This poke-yoke measure means if the wrong part is installed, the gun either won't go into battery or it'll only fire one round without cycling, because the piston isn't in contact with the op rod.

Moving on to the differences in the x39's bolt, spitting out the 7.62 x 39's heavily tapered case is a challenge for a conventional AR-style ejector. As the spent case is extracted from the chamber, the geometry of the case and chamber walls allow the case's base to pivot away from the ejector, robbing it of its full stroke and power. So SIG uses dual ejectors on the x39's boltface. They look like they line up about where a single ejector sits, but they're further toward the center than a single ejector could be. We testify to the effectiveness of the dual ejector by noting that it deposits brass in a neighboring zip code.

With regard to the x39 barrel, other than the differences necessitated by the caliber difference, it's the same as the other MCX barrels. The SBR version is an 11.5-inch barrel with a 1:9.5-inch twist rate. Thankfully, it's got the culturally correct 5/8-inch x 24 threads instead of those crazy left-handed AK threads.

The last novel area in the x39 MCX is the magazine. SIG worked with Duramag (formerly known as C-Products) to develop an enhanced magazine follower for the MCX x39. The changes come down to the geometry of the follower, how it sits in the tube, and how it acts with a load of tapered-case cartridges sitting on top of it. It prevents binding during cycling so it can keep up with the operating system, especially when *braaaaaaap* firing in a full-auto lower.

The MCX x39 was designed with parts commonality in mind. The only unswappable parts are the bolt carriage/op rod assembly, bolt, barrel, and mag.

As far as the technical mumbo jumbo, we'll call it here instead of rambling on about the rest of the MCX's common features. Well, just in case you've read this far and still aren't familiar with the MCX, know that it looks like an AR from the outside, but inside it uses a two-setting, short-stroke piston with an abbreviated bolt carriage assembly that runs without a receiver extension. It also has a quick-change barrel system that requires sliding off the handguard and torqueing just a pair of screws.

HANDLING IMPRESSIONS

The MCX includes a single-stage trigger, with our measured pull weight of 5.1 pounds. The shoe has a smooth, somewhat rolling break. There's a short take up to the wall before it continues through a non-discernable break. Overall, the trigger feels very clean with no creep.

On the range, the SBR provides a stout, thudding recoil that's tamed very well by the use of SIG's SRD762-QD silencer. In short, adding the can saw the gun undergo a character change from Rosie O'Donnell to Keanu Reeves.

RELIABILITY

SIG says its defense guns are treated to a vigorous design validation testing process to make sure these guns can run longer than any non-immortal person has a need for it to run. Beyond standard endurance testing, SIG subjects the x39 to a dry function test that it says proves the gun will run without lubrication for 600 rounds, suppressed or unsuppressed. Impressive.

We put a little more than 1,000 rounds through our MCX test gun in one afternoon. It was a long afternoon in which we almost got sick of shooting. All of our efforts to induce a malfunction were rewarded with a single failure to lock back on an empty mag.

In total, we ran seven types of ammo through the MCX over our suppressed and unsuppressed evaluation. The majority of the ammo was Brown Bear 123-grain, bimetal, polymer-coated case, FMJ. Mag dumps were the order of the day as we burned through rounds as fast as we could load the mags, slowing down only after the law enforcement class on the range next door complained about the jack-hammering noise of our unsuppressed MCX. In response, we put the SRD762-QD silencer on and, before long, they got on us again for the monotonous sound of ringing steel floating from our bay to theirs. So for the last half-dozen mags of the day, we resorted to shooting rocks and dirt balls on the berm.

Despite needing a glove to continue running the hot gun by

With the MCX x39 dressed in a side folding stock and a tiny optic combo, such as Aimpoint's ACRO P-1 in a Reptilia mount, the rifle is compact as it is reliable.

SIG SAUER MCX VIRTUS SBR

CALIBER: 7.62 x 39mm
MAGAZINE CAPACITY: 28 rounds
BARREL LENGTH: 11.5 inches
OVERALL LENGTH: 31 inches
WEIGHT: 7.4 pounds
MSRP: $2,341
URL: sigsauer.com

FEATURED ACCESSORIES:
SIG SAUER SRD762-QD silencer - $820
Aimpoint ACRO P-1 MRDS Optic - $660
Reptilia DOT Optic Mount - $95
IRS NFA Tax Stamp - $200

PRICE AS FEATURED: $4,116

AMMO	AVG. FPS	ES	SD	AVG. GROUP SIZE	BEST GROUP SIZE
BROWN BEAR 123-GRAIN BIMETAL	2,254	31	11.2	1.44 inches	1.39 inches
WOLF MILITARY CLASSIC 124-GRAIN	2,172	32	12.5	2.02 inches	1.61 inches
LAPUA 123-GRAIN	2,188	29.4	97	1.19 inches	1.10 inches
FIOCHI 124-GRAIN FMJ	1,988	73.6	201	4.97 inches	4.97 inches
AMERICAN EAGLE 124-GRAIN	1,988	11.9	47	2.28 inches	1.28 inches
SELLIER & BELLOT 123-GRAIN FMJ	2,241	12.5	41	3.01 inches	2.58 inches

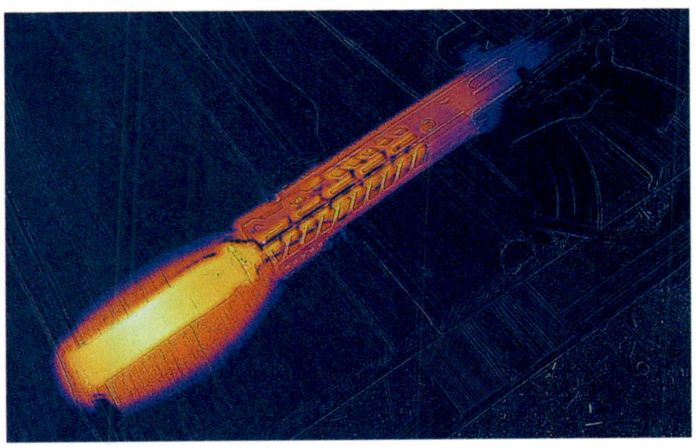

We used a Flir One camera to get an idea how hot the SBR got, and see how much heat travels from the muzzle back to the receiver. This photo was taken after continuously firing 10 mags, or about 280 rounds.

the end of the day, the MCX short-stroke piston system performed reliably. We ran another 400 rounds through the gun on a second range outing and had no malfunctions.

ACCURACY

We did our standard best of five, 5-round group, 100-yard velocity and accuracy testing with six types of ammo ... all with the SRD762-QD in place (see table above). Thanks to the effort of Lapua, we'll say the 11.5-inch MCX x39 setup approaches 1 MOA territory. We're not sure what happened with the 124-grain Fiochi, but it produced obnoxiously large groups with fear-less consistency. Everything else gave groups you'd expect.

PACKING UP

SIG says the MCX x39 lineup will be on shelves by winter 2019/2020, first as complete pistols, SBRs, and carbines (with the SBR version we tested for $2,341), then as complete upper receiver groups for around $1,400, and then possibly followed by caliber exchange kits in the $500 to $600 range based on demand. As of press time, pricing isn't set for the x39 uppers and caliber exchange kits, so SIG just offered those prices as estimates based on the cost of the current caliber exchange kits.

The 7.62 x 39 isn't the dirt-cheap, TV tube hating, all-day-blast-o-thon-at-the-quarry ammo it once was. But the ammo is readily available here again at relatively reasonable prices, and the reliable MCX x39 will hold a special place in the hearts of vehicle interdiction specialists as well as anyone needing to shoot their way out of or into a former Eastern Bloc country. R

DOWNLOAD THE FREE APP TODAY!
RECOILTV
RECOILTV
Navigate between a variety of channels
RELEVANT.
HONEST.
UNFILTERED.
RECOILTV IS THE PREMIER AUTHORITY FOR THE FIREARMS NEWS AND MEDIA YOU WANT AND DESERVE.
WWW.RECOIL.TV
RECOILTV is a 2A friendly streaming platform created to entertain and inform you on the latest guns and gear associated with the firearms lifestyle.
BROUGHT TO YOU BY:
BIG DADDY UNLIMITED
REVOLUTIONIZING THE ONLINE GUN STORE
PRESENTED BY:
BCM
EOTECH
SIGHT
norma
DEDICATED SINCE 1902
SIONYX
RECOILTV is available on:
Download on the App Store
Download on Apple TV
GET IT ON Google Play
Roku
fire tv
Chromecast